SOTHIS PRESS

Sothis Press aims to completely reimagine the public domain book with an emphasis on universal access to captivatingly designed print editions. We believe that technology will enable every single book ever published to be perpetually accessible in beautiful print form. Sothis is beauty in ubiquity.

www.sothis.press
1 Waterhouse Square, London, ECIN 2ST

The Guardians of the Columbia

Mount Hood, Mount Adams and Mount St. Helens

John H. Williams

SOTHIS PRESS

ISBN 978-1-0368-0879-2

The Guardians of the Columbia
is in the public domain.

This edition is published by
Sothis Press © 2023. The company
is registered in England and Wales.
The Company Number is 14900389.

THE GUARDIANS OF THE COLUMBIA

MOUNT HOOD, MOUNT ADAMS AND MOUNT ST. HELENS

By JOHN H. WILLIAMS
Author of "THE MOUNTAIN THAT WAS 'GOD'"

And mountains that like giants stand
To sentinel enchanted land.
SCOTT: "The Lady of the Lake."

WITH MORE THAN TWO HUNDRED ILLUSTRATIONS
INCLUDING EIGHT IN COLORS

TACOMA
JOHN H. WILLIAMS
1912

COPYRIGHT, G. M. WEISTER

Climbing the last steep slope on Mount Hood, from Cooper's Spur, With ropes anchored on summit.

Willamette River at Portland, with ships loading wheat and lumber for foreign ports.

FOREWORD

In offering this second volume of a proposed series on Western mountain scenery, I am fortunate in having a subject as unhackneyed as was that of "The Mountain that Was 'God.'" The Columbia River has been described in many publications about the Northwest, but the three fine snow-peaks guarding its great canyon have received scant attention, and that mainly from periodicals of local circulation.

These peaks are vitally a part of the vast Cascade-Columbia scene to which they give a climax. Hence the story here told by text and picture has necessarily included the stage upon which they were built up. And since the great forests of this mountain and river district are a factor of its beauty as well as its wealth, I am glad to be able to present a brief chapter about them from the competent hand of Mr. H. D. Langille, formerly of the United States forest service. A short bibliography, with notes on transportation routes, hotels, guides and other matters of interest to travelers and students, will be found at the end.

Accuracy has been my first aim. I have tried to avoid the exaggeration employed in much current writing for the supposed edification of tourists. It has seemed to me that simply and briefly to tell the truth about the fascinating Columbia country would be the best service I could render to those who love its splendid mountains and its noble river. A mass of books, government documents and scientific essays has been examined. This literature is more or less contradictory, and as I cannot hope to have avoided all errors, I shall be grateful for any correction of my text.

In choosing the illustrations, I have sought to show the individuality of each peak. Mountains, like men, wear their history on their faces,—none more so than Hood's sharp and finely scarred pyramid; or Adams, with its wide, truncated dome and deeply carved slopes; or St. Helens, newest of all our extinct volcanoes—if, indeed, it be extinct,—and least marred by the ice, its cone as perfect as Fujiyama's. Each has its own wonderful story to tell of ancient and often recent vulcanism. Let me again suggest that readers who would get the full value of the more comprehensive illustrations will find a reading glass very useful.

Thanks are due to many helpers. More than fifty photographers, professional and amateur, are named in the table of illustrations. Without their co-operation the book would have been impossible. I am also indebted for valued information and assistance to the librarians at the Portland and Tacoma public libraries, the officers and members of the several mountaineering clubs in Portland, and the passenger departments of the railways reaching that city; to Prof. Harry Fielding Reid, the eminent geologist of Johns Hopkins University;

Fred G. Plummer, geographer of the United States forest service; Dr. George Otis Smith, director of the United States geological survey; Judge Harrington Putnam, of New York, president of the American Alpine Club; Messrs. Rodney L. Glisan, William M. Ladd, H. O. Stabler, T. H. Sherrard, Judge W. B. Gilbert, H. L. Pittock, George H. Himes, John Gill, C. E. Rusk, and others in Portland and elsewhere.

The West has much besides magnificent scenery to give those who visit it. Here have been played, upon a grander stage, the closing acts in the great drama of state-building which opened three hundred years ago on the Atlantic Coast. The setting has powerfully moulded the history, and we must know one if we would understand the other. Europe, of course, offers to the American student of culture and the arts something which travel here at home cannot supply. But every influence that brings the different sections of the United States into closer touch and fuller sympathy makes for patriotism and increased national strength.

This, rather than regret for the two hundred millions of dollars which our tourists spend abroad each year, is the true basis of the "See America First" movement. According to his capacity, the tourist commonly gets value for his money, whether traveling in Europe or America. But Eastern ignorance of the West is costing the country more than the drain of tourist money.

This volume is presented, therefore, as a call to better appreciation of the splendor and worth of our own land. Its publication will be justified if it is found to merit in some degree the commendation given its predecessor by Prof. W. D. Lyman, of Whitman College, whose delightful book on the Columbia has been consulted and whose personal advice has been of great value throughout my work. "I wish to express the conviction," writes Prof. Lyman, "that you have done an inestimable service to all who love beauty, and who stand for those higher things among our possessions that cannot be measured in money, but which have an untold bearing upon the finer sensibilities of a nation."

Tacoma, June 15, 1912.

Mount Adams, seen from south slope of Mount St. Helens, near the summit, showing the Cascade ranges below. Note the great burn in the forest cover of the ridges. "Steamboat Mountain" is seen in the distance beyond. Elevation of camera, nearly 9,000 feet.

Looking up the Columbia at Lyle, Washington.

CONTENTS

ILLUSTRATIONS

The * indicates engravings from copyrighted photographs. See notice under the illustration.

THREE-COLOR HALFTONES.

ONE-COLOR HALFTONES.

MAPS.

PACIFIC OCEAN
Cape Hancock
Vancouver
Victoria
Bellingham
Mt. Baker
Everett
Seattle
Cascade Tunnel
Moclips
berdeen
So. Bend
Cathlamet
Mt. Rainier
WASHI
Castle Rock
Spirit Lake
Mt. Adams
Lake Merrill
Trout Lake
Goldendale
Hillsb
Forest Grove
McMinnville
Dallas
Oregon City
The Dalles
Mosier
Mt. Hood
Government Camp
OREG
Collville
Spokane
Coeur d'Ale
IDAH
Moscow
Walla Walla
RONDE RIVER
BLUE
La Grande
ppner

A Gray Day on the Columbia. Telephotograph of Mount Hood from the river opposite Vancouver Barracks.

Trout Lake and Mount Adams.

THE GUARDIANS OF THE COLUMBIA

I.

THE RIVER

The Columbia, viewed as one from the sea to the mountains, is like a rugged, broad-topped picturesque old oak, about six hundred miles long, and nearly a thousand miles wide, measured across the spread of its upper branches, the main limbs gnarled and swollen with lakes and lake-like expansions, while innumerable smaller lakes shine like fruit among the smaller branches.—*John Muir.*

ON a frosty morning of last July, before sunrise, I stood upon the belvedere of the delightful Cloud Cap Inn, which a public-spirited man of Portland has provided for visitors to the north side of Mount Hood; and from that superb view-point, six thousand feet above sea level, watched the day come up out of the delicate saffron east. Behind us lay Eliot Glacier, sloping to the summit of the kindling peak. Before us rose—an ocean!

Mount St. Helens, seen from the Columbia at Vancouver, with railway bridge in foreground.

Never was a marine picture of greater

COPYRIGHT, G. M. WEISTER

View up the Columbia on north side, opposite Astoria. Noon rest of the night fishermen. Much of the fishing on the lower Columbia is done at night with gill-nets from small boats. The river is here six miles wide.

stress. No watcher from the crags, none who go down to the sea in ships, ever beheld a scene more awful. Ceaselessly the mighty surges piled up against the ridge at our feet, as if to tear away the solid foundations of the mountain. Towers and castles of foam were built up, huge and white, against the sullen sky, only to hurl themselves into the gulf. Far to the north, dimly above this gray and heaving surface were seen the crests of three snow-mantled mountains, paler even than the undulating expanse from which they emerged. All between was a wild sea that rolled across sixty miles of space to assail those ghostly islands.

Yet the tossing breakers gave forth no roar. It was a spectral and pantomimic ocean. We "had sight of Proteus rising from the sea," but no Triton of the upper air blew his "wreathed horn." Cold and uncanny, all that seething ocean was silent as a windless lake under summer stars. It was a sea of clouds.

Swiftly the dawn marched westward. The sun, breaking across the eastern ridges, sent long level beams to sprinkle the cloud-sea with silver. Its touch was magical. The billows broke and parted. The mists fled in panic. Cloud after cloud arose and was caught

Astoria in 1813, showing the trading post established by John Jacob Astor.

Looking north from lower end of Eliot Glacier on Mount Hood, across the Cascade ranges and the Columbia River canyon, twenty-five miles away, to Mount Adams (right), Mount Rainier-Tacoma (center), and Mount St. Helens (left). These snow-peaks are respectively 60, 100, and 60 miles distant.

away into space. The tops of the Cascade ranges below came, one by one, into view. Lower and lower, with the shortening shadows, the wooded slopes were revealed in the morning light. Here and there some deep vale was still white and hidden. Scattered cloud-fleeces clung to pinnacles on the cliffs. Northward, the snow-peaks in Washington towered higher. Great banks of fog embraced their forested abutments, and surged up to their glaciers. But the icy summits smiled in the gladness of a new day. The reign of darkness and mist was broken.

> Never did sun more beautifully steep
> In his first splendor valley, rock or hill.

Clearer and wider the picture grew. Below us, the orchards of Hood River caught the fresh breezes and laughed in the first sunshine. The day reached down into the nearer canyons, and saluted the busy, leaping brooks. Noisy waterfalls filled the glens with spray, and built rainbows from bank to bank, then hurried and tumbled on, in conceited haste, as if the ocean must run dry unless replenished by their wetness ere the sun should set again. Rippling lakes, in little mountain pockets, signaled their joy as blankets of dense vapor were folded up and quickly whisked away.

Columbia Slough in Winter, near the mouth of the Willamette.

Thirty miles northeast, a ribbon of gold flashed the story of a mighty stream at The Dalles. Far beyond, even to the uplands of the Umatilla and the Snake, to the Blue Mountains of eastern Washington and Oregon, stretched the wheat fields and stock ranges of that vast "Inland Empire" which the great river watered; while westward, cut deep

Cape Horn, tall basaltic cliffs that rise, terrace upon terrace, on the north side of the Columbia, twenty-five miles east of Portland. Lone Rock is seen in the distance.

COPYRIGHT, G. M. WEISTER

St. Peter's Dome, an 800-foot crag on the south bank of the Columbia; Mt. Adams in the distance

'Uplift against the blue walls of the sky
Your mighty shapes, and let the sunshine weave
ts golden network in your belting woods;
Smile down in rainbows from your falling floods,

Mount Hood, seen from Columbia Slough.

through a dozen folds of the Cascades, the chasm it had torn on its way to the sea was traced in the faint blue that distance paints upon evergreen hills. Out on our left, beyond the mountains, the Willamette slipped down its famous valley to join the larger river; and still farther, a hundred and fifty miles away, our glasses caught the vague gray line of the Pacific. Within

Campfire of Yakima Indians gathered at the Astoria Centennial, 1911, to take part in "The Bridge of the Gods," a dramatization of Balch's famous story. The celebration of the one hundredth anniversary of the Astor trading post at the mouth of the Columbia was made noteworthy by a revival of Indian folk lore, in which the myth of the great tamahnawas bridge held first place.

Sunset at the mouth of the Columbia. Cape Hancock on right, Point Adams on left. View from river off Astoria.

Northern part of Portland, showing the Willamette River flowing through it, and indicating relative position of the three snow-peaks. Mount Hood (right) and Mount St. Helens (left) are each about fifty miles away, while Mount Adams, seen between, is twenty miles farther.

these limits of vision lay a noble and historic country, the lower watershed of the Columbia.

Earth has not anything
to show more fair.

Wide as was the prospect, however, it called the imagination to a still broader view; to look back, indeed,—how many millions of years?—to an earlier dawn, bounded by the horizons of geological time. Let us try to realize the panorama thus unfolded. As we look down from some aerial viewpoint,

"The Coming of the White Man" and "SacajaWea," statues in Portland City Park Which commemorate the aboriginal Americans.

behold! there is no Mount Hood and no Cascade Range. The volcanic snow-peaks of Oregon and Washington are still embryo in the womb of earth. We stand face to face with the beginnings of the Northwest.

Far south and east of our castle-in-the-air, islands rise slowly out of a Pacific that has long rolled, unbroken, to the Rocky Mountains.

Sunset on Vancouver Lake, near Vancouver, Washington.

We see the ocean bed pushed above the tide in what men of later ages will call the Siskiyou and the Blue Mountains, one range in southwestern, the other in eastern, Oregon. A third uptilt, the great Okanogan, in northern Washington, soon appears. All else is sea. Upon these primitive uplands, the date is written in the fossil archives of their ancient sea beaches, raised thousands of feet above the former shore-line level. At a time when all western Europe was still ocean, and busy foraminifers were strewing its floor with shells to form the chalk beds of France and England, these first lands of our Northwest emerged from the great deep. It is but a glimpse we get into the immeasurable distance of the Paleozoic. Its time-units are centuries instead of minutes.

Another glance, as the next long geological age passes, and we perceive a second step in the making of the West. It is the gradual uplift of a thin sea-dike, separating the two islands first disclosed, and stretching from the present Lower California to our Alaska. It is a folding of the earth's crust that will, for innumerable ages, exercise a controlling influence upon the whole western slope of North America. At first merely a sea-dike, we see it slowly become a far-reaching range of hills, and then a vast continental mountain

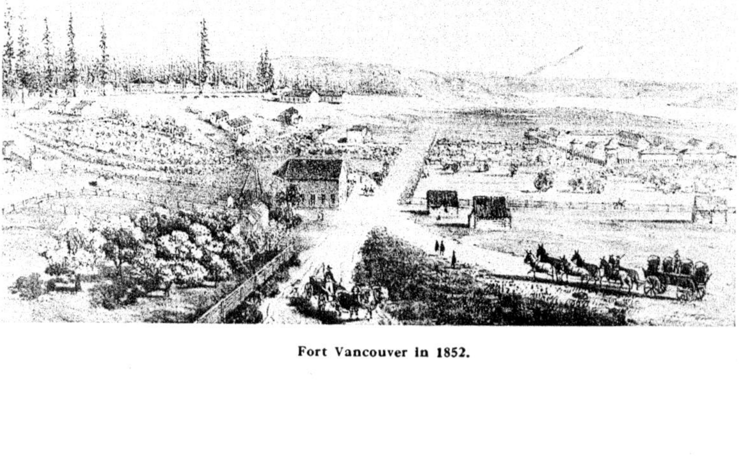

Fort Vancouver in 1852.

COPYRIGHT, G. M. WEISTER

Rooster Rock, south bank of the Columbia.

system, covering a broad region with its spurs and interlying plateaus. "The highest mountains," our school geographies used to tell us, "parallel the deepest oceans." So here, bordering its profound depths, the Pacific ocean, through centuries of centuries, thrust upward, fold on fold, the lofty ridges of this colossal Sierra-Cascade barrier, to be itself a guide of further land building, a governor of climate, and a reservoir of water for valleys and river basins as yet unborn.

Behind this barrier, what revolutions are recorded! The inland sea, at first a huge body of ocean waters, becomes in time a fresh-water lake. In its three thousand feet of sediment, it buries the fossils of a strange reptilian life, covering hundreds of thousands of years. Cycle follows cycle, altering the face of all that interior basin. Its vast lake is lessened in area as it is cut off from the Utah lake on the south and hemmed in

Seining for salmon on the lower Columbia.

COPYRIGHT, KISER PHOTO CO.

The Columbia near Butler, looking across to Multnomah Falls.

by upfolds on the north. Then its bed is lifted up and broken by forces of which our present-day experiences give us no example. Instead of one great lake, as drainage proceeds, we behold at last a wide country of many lakes and rivers. Their shores are clothed in tropical vegetation. Under the palms, flourish a race of giant mammals. The broad-faced ox, the mylodon, mammoth, elephant, rhinoceros, and mastodon, and with them the camel and the three-toed horse, roam the forests that are building the coal deposits for a later age. This story of the Eocene and Miocene time is also told in the fossils of the period, and we may read it in the strata deposited by the lakes.

Captain Som-Kin, chief of Indian police, Umatilla reservation.

Age succeeds age, not always distinct, but often overlapping one another, and all changing the face of nature. The Coast Range rises, shutting in vast gulfs to fill later, and form the valleys of the Sacramento

and San Joaquin in California and the Willamette in Oregon, with the partly filled basin of Puget Sound in Washington. Centering along the Cascade barrier, an era of terrific violence shakes the very foundation of the Northwest. Elevations and contours are changed. New lake beds are created. Watersheds and stream courses are remodeled. Dry "coulees" are left where formerly

Multnomah Falls in Summer and Winter. This fascinating cascade, the most famous in the NorthWest, falls 720 feet into a basin, and then 130 feet to the bank of the Columbia below.

rivers flowed. Strata are uptilted and riven, to be cross-sectioned again by the new rivers as they cut new canyons in draining the new lakes. Most important of all, outflows of melted rock, pouring from fissures in the changing earth-folds, spread vast sheets of basalt, trap and andesite over most of the interior. Innumerable craters build cones of lava and scoriæ along the Cascade uptilt, and scatter clouds of volcanic ashes upon the steady sea winds, to blanket the country for hundreds of miles with deep layers of future soil.

A reign of ice follows the era of tropic heat. Stupendous glaciers grind the volcanic rocks, and carving new valleys, endow them with fertility for new forests that will rise where once the palm forests stood. With advancing age, the earth grows cold and quiet, awakening only to an occasional volcanic eruption or earthquake as a reminder of former violence. The dawn of history approaches. The country slowly takes on its present shape. Landscape

View from the cliffs at Multnomah Falls (seen on right). Castle Rock is in distance on north side.

changes are henceforth the work of milder forces, erosion by streams and remnant glaciers. Man appears.

Throughout the cycles of convulsion and revolution which we have witnessed from our eyrie in the clouds, the vital and increasing influence in the building of the Northwest has been the Cascade upfold. First, it merely shuts in a piece of the Pacific. Rising higher, its condensation of the moist ocean wind feeds the thousand streams that convert the inland seas thus enclosed from salt to fresh water, and furnish the silt deposited over their floors. The fractures and faults resulting from its uptilting spread an empire with some of the largest lava flows in geological history. It pushes its snow-covered volcanoes

The broad Columbia, seen from Lone Rock, a small island east of Cape Horn. Shows successive ranges of the Cascades cut by the river, with Archer and Arrowhead Mountains and Castle Rock in distance on north side.

Castle Rock, a huge tower of columnar basalt, 1146 feet high, on north bank of the Columbia, forty miles east of Portland. View from Mosquito Island.

COPYRIGHT, KISER PHOTO CO.

The Columbia, opposite Oneonta Bluffs and Gorge, and Horsetail Falls.

upward, to scatter ashes far to the east. Finally, its increasing height converts a realm of tropical verdure into semi-arid land, which only its rivers, impounded by man, will again make fertile.

An original American — "Jake" Hunt, former Klickitat chief, 112 years old. He is said to be the oldest Indian on the Columbia.

In all this great continental barrier, throughout the changes which we have witnessed, there has been only one sea-level pass. For nearly a thousand miles northward from the Gulf of California, the single outlet for the waters of the interior is the remarkable canyon which we first saw from the distant roof of Cloud Cap Inn. Here the Columbia, greatest of Western rivers, has cut its way through ranges rising more than 4,000 feet on either hand. This erosion, let us remember, has been continuous and gradual, rather than the work of any single epoch. It doubtless began when the Cascade Mountains were in their infancy, a gap in the prolonged but low sea-dike. The drainage, first of the vast salt lake shut off from the ocean, and then of the succeeding fresh-water lakes, has preserved this channel to the sea, cutting it deeper and deeper as the earth-folds rose higher, until at last the canyon became one of the most important river gorges in the world. Thus nature prepared a vast and fruitful section of the continent for human

View from 2,300 foot elevation, west of St. Peter's Dome. The Columbia here hurries down from The Cascades with a speed varying in different seasons from six to ten miles per hour. Mosquito Island lies below, with Castle Rock opposite. Beyond, the beautiful wooded ridges rise to 4,100 feet in Arrowhead and Table Mountains, and the snowy dome of Mount Adams closes the scene, fifty miles away.

COPYRIGHT, G. M. WEISTER

Oneonta Gorge, south side of the Columbia, thirty-three miles east of Portland.

use, and provided it with a worthy highway to the ocean.

Over this beautiful region we may descry yet another dawn, the beginnings of the Northwestern world according to Indian legend. The Columbia River Indian, like his brothers in other parts of the country, was curious about the origin of the things he beheld around him, and oppressed by things he could not see. The mysteries both of creation and of human destiny weighed heavily upon his blindness; and his mind, pathetically groping in the dark, was ever seeking to penetrate the distant past and the dim future. So far as he had any religion, it was connected with the symbols of power in nature, the forces which he saw at work about him. These forces were often terrible and ruinous, so his gods were as often his enemies as his benefactors. Feeling his powerlessness against their cunning, he borrowed a cue from the "animal people," Watetash, who used craft to circumvent the malevolent gods.

These animal people, the Indian believed, had inhabited the world before the time of the first grandfather, when the sun was as yet only a star, and the earth, too, had grown but little, and was only a small island. The chief of the animal people was Speelyei, the coyote, not the mightiest but the shrewdest of them all. Speelyei was the friend of "people". He had bidden people to appear, and they "came out."

One of the most interesting attempts to account for the existence of the Red Man in the Northwest is the Okanogan legend that tells of an island

Looking up the Columbia, near Bonneville. The main channel of the river is on right of the shoal in foreground.

far out at sea inhabited by a race of giant whites, whose chief was a tall and powerful woman, Scomalt. When her giants warred among themselves, Scomalt grew angry and drove all the fighters to the end of the island. Then she broke off the end of the island, and pushing with her foot sent it floating away over the sea. The new island drifted far. All the people on it died save one man and one woman. They caught a whale, and its blubber saved them from starving. At last they escaped from the island by making a canoe. In this they paddled many days. Then they came to the mainland, but it was small. It had not yet grown much. Here they landed. But while they had been in the canoe, the sun had turned them from white to red. All the Okanogans were their children. Hence they all are red. Many years from now the whole of the mainland will be cut loose from its foundations,

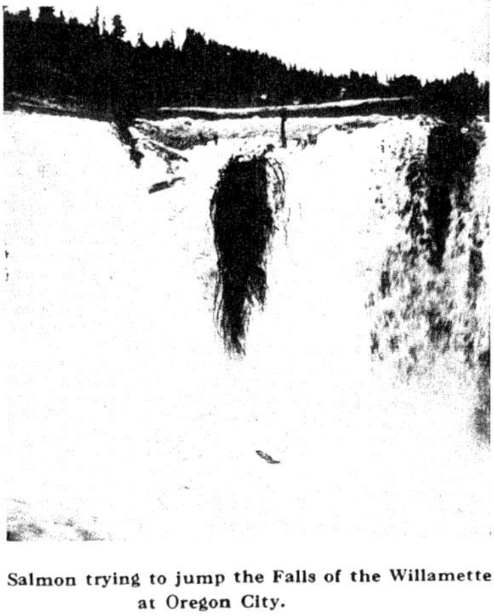

Salmon trying to jump the Falls of the Willamette at Oregon City.

In the Columbia Canyon at Cascade, with train on the "North Bank" road.

The Cascades of the Columbia. The narrow, rock-filled channel has a fall of thirty-seven feet in four miles. Here the river meets the tides from the ocean, 160 miles away. On the opposite bank, at right, is seen Table Mountain, 4,100 feet, the north abutment of the legendary "Bridge of the Gods."

COPYRIGHT, G. M. WEISTER

Fishwheel below the Cascades, with Table Mountain on north side of river.

and become an island. It will float about on the sea. That will be the end of the world.

To the aboriginal Americans in the Northwest the great river, "Wauna" in their vocabulary, was inevitably a subject of deep interest. It not only furnished them a highway, but it supplied them with food. Their most fascinating myths are woven about its history. One of these told of the mighty struggle between Speelyei and Wishpoosh, the greedy king beaver, which resulted in breaking down the walls of the great lakes of the interior and creating a passage for their waters through the mountains. Thus the Indians accounted for the Columbia and its canyon.

But first among the river myths must always be the Klickitat legend of the famous natural bridge, fabled to have stood where the Cascades of the

COPYRIGHT, KISER PHOTO CO.

Sunrise on the Columbia; view at 4 a. m. from top of Table Mountain.

Nightfall on the Columbia.

"O love, they die in yon rich sky,
    They faint on hill or field or river:
Our echoes roll from soul to soul,
    And grow forever and forever."—Tennyson.

Looking down the Columbia below the Cascades, showing many ranges cut by the river. On the left of the scene is "Sliding Mountain," its name a reminder that the hillsides on both banks are slowly moving toward the stream and compelling the railways occasionally to readjust their tracks.

Wind Mountain and remnant of submerged forest, above the Cascades, at low Water.

Columbia now are. This is one of the most beautiful legends connected with the source of fire, a problem of life in all the northern lands. Further, it tells the origin of the three snow-peaks that are the subject of this book.

In the time of their remote grandfathers, said the Klickitats, Tyhee Saghalie, chief of the gods, had two sons. They made a trip together down the river to where The Dalles are now. The sons saw that the country was beautiful, and quarrelled as to its possession. Then Saghalie shot an arrow to the north and an arrow to the west. The sons were bidden to find the arrows, and settle where they had fallen. Thus one son settled in the fair country between the great river and the Yakima, and became the grandfather of the Klickitats. The other son settled in the Willamette valley and became the ancestor of the large Multnomah tribe. To keep peace between

Steamboat entering Cascade Locks.

Moonlight upon the Columbia, with clouds on Wind Mountain. Looking up the river from the Cascades.

the two tribes, Saghalie raised the great mountains that separate those regions. But there were not yet any snow-peaks. The great river also flowed very deep between the country of the Klickitats and the country of the Multnomahs. That the tribes might always be friendly, Saghalie built a huge bridge of stone over the river. The Indians called it the tamahnawas bridge, or bridge of the gods. The great river flowed under it, and a witch-woman,

White Salmon River and its Gorge, south of Mount Adams.

Loowit, lived on it. Loowit had charge of the only fire in the world.

Loowit saw how miserable the tribes were without fire. Therefore she besought Saghalie to permit her to give them fire. Saghalie granted her request. Thus a fire was kindled on the bridge. The Indians came there and obtained fire, which greatly improved their condition. Saghalie was so much pleased

Looking down the Columbia Canyon from the cliffs at White Salmon, Washington.

with Loowit's faithfulness that he promised the witch-woman anything she might ask. Loowit asked for youth and beauty. So Saghalie transformed her into a beautiful maiden.

Many chiefs fell in love with Loowit because of her beauty. But she paid heed to none till there came two other chiefs, Klickitat from the north, Wiyeast from the west. As she could not decide which of them to accept as her husband, they and their people went to war. Great distress came upon the people because of this fighting. Saghalie grew angry at their evil doing, and determined to punish them. He broke down the tamahnawas bridge, and put Loowit,

An Oregon Trout Stream.

Looking up the Columbia from Hood River, Oregon.

Wiyeast and Klickitat to death. But they had been beautiful in life, therefore Saghalie would have them beautiful in death. So he made of them the three famous snow-peaks. Wiyeast became the mountain which white men call Mount Hood; Klickitat became Mount Adams; Loowit was changed into Mount St. Helens. Always, said Saghalie, they should be clothed in garments of snow.

Thus was the wonderful tamahnawas bridge destroyed, and the great river dammed by the huge rocks that fell into it. That caused the Cascade rapids. Above the rapids, when the river is low, you can still see the forests that were buried when the bridge fell down and dammed the waters.

This noteworthy myth, fit to rank with the folk-lore masterpieces of any primitive people, Greek or Gothic, is of course only a legend. The

COPYRIGHT, B. A. GIFFORD

Hood River, fed by the glaciers of Mount Hood.

A Late Winter Afternoon. View across the Columbia from White Salmon to the mouth of Hood River, showing the Hood River Valley with Mount Hood wrapped in clouds.

Indian was not a geologist. True, we see the submerged forests to-day, at low water. But their slowly decaying trunks were killed, perhaps not much more than a century ago, by a rise in the river that was not caused by the fall of a natural bridge, but by a landslide from the mountains.

There is a slow and glacier-like motion of the hillsides here which from time to time compels the railways on either bank to readjust their tracks. The rapids at the Cascades, with their fall of nearly forty feet, are doubtless the result of comparatively recent volcanic action. Shaking down vast masses of rock, this dammed

Memaloose Island, or Island of the Dead, last resting place of thousands of Indians. The lone monument is that of Maj. Victor Trevitt, a celebrated pioneer, who asked to be buried here among "honest men."

"Gateway to the Inland Empire." Towering cliffs of stratified lava that guard the Columbia on each bank at Lyle, Washington.

the river, and caused it to overflow its wooded shores above. But to the traveler on a steamboat breasting the terrific current below the government locks, as he looks up to the towering heights on either side of the narrowed channel, the invention of poor Lo's untutored mind seems almost as easy to believe as the simpler explanation of the scientist.

Remarkable as is this fire myth of the tamahnawas bridge, the legend inspired by the peculiarities of northwestern climate is no less beautiful. This climate differs materially, it is well known, from that of eastern America in the same latitude. The Japan Current warms the coast of Oregon and Washington just as the Gulf Stream warms the coast of Ireland. East of the

"Grant Castle" and Palisades of the Columbia, on north side of the river below The Dalles.

Cascade Mountains, the severe cold of a northern winter is tempered by the "Chinook" winds from the Pacific. A period of freezing weather is shortly followed by the melting of the snow upon the distant mountains; by night the warm Chinook sweeps up the Columbia canyon and across the passes, and in a few hours the mildness of spring covers the land.

Such a phenomenon inevitably stirred the Indian to an attempt to interpret it. Like the ancients of other races, he personified the winds. The Yakima account of the struggle between the warm winds from the coast and the icy blasts out of the Northeast will bear comparison with the Homeric

COPYRIGHT, G. M. WEISTER

The Dalles of the Columbia, lower channel, east of Dalles City. The river, crowded into a narrow flume, flows here at a speed often exceeding ten miles an hour.

tale of Ulysses, buffeted by the breezes from the bag given him by the wind-god Aeolus.

Five Chinook brothers, said the Yakima tradition, lived on the great river. They caused the warm winds to blow. Five other brothers lived at Walla Walla, the meeting place of the waters. They caused the cold winds. The grandparents of them all lived at Umatilla, home of the wind-blown sands. Always there was war between them. They swept over the country, destroying the forests, covering the rivers with ice, or melting the snows and causing floods. The people suffered much because of their violence.

Cabbage Rock, a huge freak of nature standing in the open plain four miles north of The Dalles. Apparently, the lava core of a small extinct crater.

Then Walla Walla brothers challenged Chinook brothers to wrestle. Speelyei, the coyote god, should judge the contest. He should cut

A True Fish Story of the Columbia, Where four- and even five-foot salmon are not uncommon.

off the heads of those who fell.

The crafty Speelyei secretly advised the grandparents of Chinook brothers that if they would throw oil on the ground, their sons would not fall. This they did. But Speelyei also told the grandparents of Walla Walla brothers that if they would throw ice on the ground, their sons would not fall. This they did. So the Chinook brothers were thrown one after another, and Speelyei cut off their heads, according to the bargain. So the five Chinook brothers were dead.

But the oldest of them left an infant son. The child's mother brought him up to avenge the killing of his kinsmen. So the son grew very strong, until he could pull up great fir trees as if they were weeds. Then Walla Walla brothers challenged Young Chinook to wrestle. Speelyei should judge the contest. He should cut off the heads of those who fell. Secretly Speelyei advised Young Chinook's grandparents to throw oil on the ground last. This they did. So Walla Walla brothers were thrown one after another by Young Chinook, until four of them had fallen. Only the youngest of them was left. His heart failed him, and he refused to wrestle. Speelyei pronounced this sentence upon him: "You shall live, but you shall no longer have power to freeze people." To Young Chinook, he said: "You must blow only lightly, and you must blow first upon the mountains, to warn people of your coming."

The last dawn of all opens upon the white man's era. On the Columbia, recorded history is recent, but already

The Zig-zag river in Winter, south side of Mount Hood.

The Dalles. This name, meaning literally flat stones, was given by the early French-Canadian voyageurs to the twelve-mile section below Celilo, where, the Columbia has cut through the level lava strata, forming a channel in some places less than 200 feet wide and nearly 200 feet deep at low water. At higher stages the river fills many lateral channels and roars past many islands of its own carving.

The "Witch's Head," an Indian picture rock at the old native village of Wishram, north side of the Columbia near Celilo Falls. The Indians believe that if an unfaithful wife passes this rock, its eyes follow her with mute accusation.

epic. Its story is outside the purpose of this volume. But it is worth while, in closing our brief glance at the field, to note that this story has been true to its setting. Rich in heroism and romance, it is perhaps the most typical, as it is the latest, chapter in the development of the West. For this land of the river, its quarter-million square miles stretching far northward to Canada, and far eastward to the Yellowstone, built about with colossal mountains, laced with splendid waterways, jeweled with beautiful lakes, where upheaval and eruption, earthquake and glacier have prepared a home for a great and happy population, has already been the scene of a drama of curious political contradictions and remarkable popular achievement.

The Columbia River basin, alone of all the territories which the United States has added to its original area, was neither bought with money nor annexed by war. Its acquisition was a triumph of the American pioneer. Many nations looked with longing to this Northwest, but it fell a prize to the nation that neglected it. Spain and Russia wished to own it. Great

Village of Indian Tepees, Umatilla Reservation, near Pendleton, Oregon. Many of these Indians are rich landoWners, but they prefer tents to houses.

Mount Adams, seen from Eagle Peak in the Rainier National Park. View shows some of the largest earth-folds in the Cascade Range, with the great canyon of the Cowlitz, one of the tributaries of the Columbia River. Elevation of camera 6,000 feet.

Britain claimed and practically held it. The United States ignored it. For nearly half a century after the discovery of the river by a Yankee ship captain, Robert Gray, in 1792, and its exploration by Jefferson's expedition under Lewis and Clark, in 1805, its ownership was in question. For several decades after an American merchant, John Jacob Astor, had established the first unsuccessful trading post, in 1811, the country was actually ruled by the British through a private corporation. The magic circle drawn about it by the Hudson's Bay Company seemed impenetrable. Held nominally by the American and British governments in joint occupancy, it was in fact left to the halfbreed servants of a foreign monoply that sought to hold an empire for its fur trade, and to exclude settlers because their farms would interfere with its beaver traps. Congress deemed the region worthless.

A clearing in the forest. Mount Hood from Sandy, twenty-five miles west of the peak.

But while sleepy diplomacy played its game of chess between

An Indian Madonna and Child. Umatilla Reservation.

Washington and London, the issue was joined, the title cleared and possession taken by a breed of men to whom the United States owes more than it can ever pay. From far east came the thin vanguard of civilization which, for a century after the old French and Indian war, pushed our boundaries resistlessly westward. It had seized the "dark and bloody ground" of Kentucky. It had held the Ohio valley for the young republic during the Revolution. It had built states from the Alleghanies to the Mississippi. And now, dragging its wagons across the plains and mountains, it burst, sun-browned and half-starved, into Oregon. Missionaries and traders, farmers, politicians and speculators, it was part of that army of restless spirits who, always seeing visions of more fertile lands and rising cities beyond, stayed and long in no place, until at last they found their way barred by the Pacific, and therefore stayed to build the commonwealths of Oregon, Washington and Idaho.

The arena of their peaceful contest was worthy of their daring. " 'A land of old upheaven from the abyss,' a land of deepest deeps and highest heights, of richest verdure here, and barest desolation there, of dense forest on one side, and wide extended prairies on the other; a land of contrasts, contrasts in contour, hues, productions, and history,"—thus Professor Lyman describes the stage which the pioneers found set for them.

Finished portion of Canal at Celilo, Which the Government is building around TumWater Falls and The Dalles.

The grim sentinels of "the Wallula Gateway," huge basaltic pillars that rise on the south bank of the river, where it crosses the Washington-Oregon line. View looking south.

COPYRIGHT, KISER PHOTO CO.

TumWater, the falls of the Columbia at Celilo; total drop, tWenty feet at low Water. In Summer, When the snoW on the Bitter Root and Rocky Mountains is melting, the river rises often more than sixty feet. Steamboats have then passed safely doWn. Wishram, an ancient Indian fishing village, was on the north bank beloW the falls, and Indians may often still be seen spearing salmon from the shores and islands here.

The tremendous problems of its development, due to its topography, its remoteness, its magnificent distances, and its lack of transportation, demanded men of sturdiest fiber and intrepid leading. No pages of our history tell a finer story of action and initiative than those which enroll the names of McLoughlin, the great Company's autocratic governor, not unfitly called "the father of Oregon," and Whitman, the martyr, with the frontier leaders who fashioned the first ship of state launched in the Northwest, and their contemporaries, the men who built the first towns, roads, schools, mills, steamboats and railways.

Macaulay tells us that a people who are not proud of their forebears will never deserve the pride of their descendents. The makers of Old Oregon included as fair a proportion of patriots and heroes as the immigrants of the Mayflower. We who journey up or down the Columbia in a luxurious steamer, or ride in a train *de luxe* along its banks, are the heirs of their achievement. Honor to the dirt-tanned ox-drivers who seized for themselves and us this empire of the river and its guardian snow-peaks!

A lordly river, broad and deep,
With mountains for its neighbors, and in view
Of distant mountains and their snowy tops.

Summit of Mount Hood, viewed from western end of the ridge, showing north side of the peak in July.

Columbia River and Mt. Hood, seen from White Salmon, Washington.

"Beloved mountain, I
Thy worshiper, as thou the sun's, each morn
My dawn, before the dawn, receive from thee;
And think, as thy rose-tinted peak I see,
That thou wert great when Homer was not born,
And ere thou change all human song shall die."—Helen Hunt Jackson

North side of Mount Hood, from ridge several miles West of Cloud Cap Inn. View shows gorges cut by the glacier-fed streams. Cooper Spur is on left sky line. Barret Spur is the great ridge on right, with Ladd glacier canyon beyond. Coe glacier is in center.

II.

THE MOUNTAINS.

Silent and calm, have you e'er scaled the height
Of some lone mountain peak, in heaven's sight?
—*Victor Hugo.*

There stood Mount Hood in all the glory of the alpen glow, looming immensely high, beaming with intelligence. It seemed neither near nor far. * * * The whole mountain appeared as one glorious manifestation of divine power, enthusiastic and benevolent, glowing like a countenance with ineffable repose and beauty, before which we could only gaze with devout and lowly admiration.—*John Muir.*

FROM the heights which back the city of Portland on the west, one may have a view that is justly famous among the fairest prospects in America. Below him lies the restless city, busy with its commerce. Winding up from the south comes the Willamette, its fine valley narrowed here by the hills, where the river forms Portland's harbor, and is lined on either side with mills and shipping. Ten miles beyond, the Columbia flows down from its canyon on the east, and turns north-

Winter on Mount Hood. The roof of the club house of the Portland Snow-shoe Club is seen over the ridge.

Watching the climbers from the plaza at Cloud Cap Inn, northeast side of Mount Hood. Immediately in front, Eliot glacier is seen, dropping into its canyon on the right. On the left is Cooper Spur, from which a sharp ascent leads to the summit of the peak.

ward, an expanding waterway for great vessels, to its broad pass through the Coast Range. In every direction, city and country, farm and forest, valley and mountain, stretches a noble perspective. From the wide rivers and their shining borders, almost at sea level, the scene arises, terrace upon terrace, to the encircling hills, and spreads across range after range to the summits of the great Cascades.

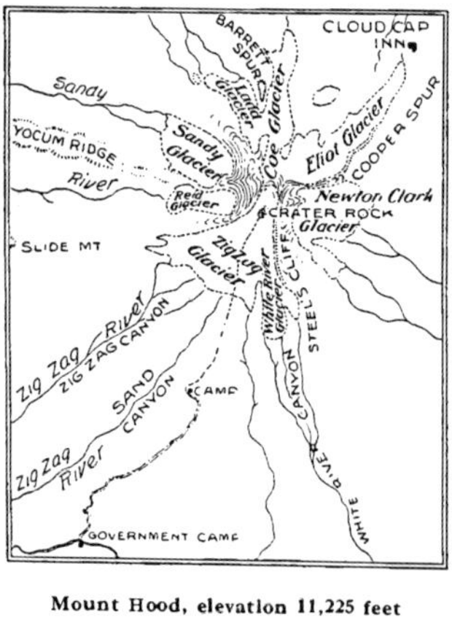

Mount Hood, elevation 11,225 feet

Dominating all are the snow-peaks, august sentinels upon the horizon. On a clear day, the long line of them begins far down in central Oregon, and numbers six snowy domes. But any average day includes in its glory the three nearest, Hood, Adams, and St. Helens. Spirit-like, they loom above the soft Oregon haze, their glaciers signaling from peak to peak, and their shining summits bidding the sordid world below to look upward.

Nature has painted canvases more colorful, but none more perfect in its strength and rest. Here is no flare of the desert, none of the flamboyant, terrible beauty of the Grand Canyon. It is a land

Lower end of Eliot glacier, seen from Cooper Spur, and showing the lateral moraines which this receding glacier has built in recent years.

of warm ocean winds and cherishing sunshine, where the emeralds and jades of the valleys quickly give place to the bluer greens of evergreen forests that cover the hill country; and these, in turn, as distance grows, shade into the lavenders and grays of the successive ranges. The white peaks complete the picture with its most characteristic note. They give it distinction.

Such a panorama justifies Ruskin's bold assertion: "Mountains are the beginning and end of all natural scenery." Without its mountains, the view from Council Crest would be as uninteresting as that from any tower in any prairie

Snout of Eliot glacier, its V-shaped ice front heavily covered with morainal debris.

Cone of Mount Hood, seen from Cooper Spur on northwest side. A popular route to the summit leads along this ridge of volcanic scoriae and up the steep snow slope above.

city. But all mountains are not alike. In beginning our journey to the three great snow-peaks which we have viewed from Portland heights, it is well to define, if we may, the special character of our Northwestern scene. We sometimes hear the Cascade district praised as "the American Switzerland." Such a comparison does injustice alike to our mountains and to the Alps. As a wild, magnificent sea of ice-covered mountain tops, the Alps have no parallel in America. As a far-reaching system of splendid lofty ranges clothed in the green of dense forests and surmounted by towering, isolated summits of snowy volcanoes, the Cascades are wholly without their equal in Europe. This is the testimony of famous travelers and alpinists, among them Ambassador Bryce, who has written of our Northwestern mountain scenery:

We have nothing more beautiful in Switzerland or Tyrol, in Norway or in the Pyrenees. The combination of ice scenery with woodland scenery of the grandest type is to be found nowhere in the Old World, unless it be in the Himalayas, and, so far as we know, nowhere else on the American continent.

In his celebrated chapter of the "Modern Painters" which describes the sculpture of the mountains, Ruskin draws a picture of the Alps that at once sets them apart from the Cascades:

The longer I stayed among the Alps, the more I was struck by their being a vast plateau, upon which nearly all the highest peaks stood like children set upon a table, removed far back from the edge, as if for fear of their falling. The most majestic scenes are produced by one of the great peaks having apparently walked to the edge of the table to look over, and thus showing itself sud-

Cloud Cap Inn, north side of Mount Hood. Elevation 5,900 feet.

Portland's White Sentinel, Mount Hood. Telephoto view from City Park, showing a portion of the city, with modern buildings and smoke of factories.

COPYRIGHT, G. M. WEISTER

Ice cascade on Eliot glacier, Mount Hood.

denly above the valley in its full height. But the raised table is always intelligibly in existence, even in these exceptional cases; and for the most part, the great peaks are not allowed to come to the edge of it, but remain far withdrawn, surrounded by comparatively level fields of mountain, over which the lapping sheets of glacier writhe and flow. The result is the division of Switzerland into an upper and lower mountain world; the lower world consisting of rich valleys, the upper world, reached after the first steep banks of 3,000 to 4,000 feet have been surmounted, consisting of comparatively level but most desolate tracts, half covered by glacier, and stretching to the feet of the true pinnacles of the chain.

Portland Snow-shoe Club members on Eliot glacier in Winter.

Nothing of this in the Cascades! Instead, we have fold upon fold of the earth-crust, separated by valleys of great depth. The ranges rise from levels but little above the sea. For example, between Portland and Umatilla, although they are separated by the mountains of greatest actual elevation in the United States, there is a difference of less than two hundred and fifty feet, Umatilla, east of the Cascades, being only two hundred and ninety-four feet above tide. Trout Lake, lying below Mount Adams, at the head of one of the great intermountain valleys, has an elevation of less than two thousand feet.

Thus, instead of the Northwestern snow-peaks being set far back upon a general upland and hid-

COPYRIGHT, G. M. WEISTER

Snow-bridge over great crevasse, near head of Eliot glacier.

den away behind lesser mountains, to be seen only after one has reached the plateau, thousands of feet above sea level, they actually rise either from comparatively low peneplanes on one side of the Cascades, as in the case of St. Helens, or from the summit of one of the narrow, lofty ridges, as do Hood and Adams. But in either case, the full elevation is seen near at hand and from many directions—an elevation, therefore, greater and more impressive than that of most of the celebrated Alpine summits.

Famous as is the valley of Chamonix, and noteworthy as are the glaciers to which it gives close access, its views of Mont Blanc are disappointing. Not until the visitor has scaled one of the neighboring *aiguilles*, can he command a satisfactory outlook toward the Monarch of the Alps. And nowhere in Switzerland do I recall a picture of such memorable splendor as greets the traveler from the Columbia, journeying either southward, up the Hood River Val-

COPYRIGHT, G. M. WEISTER

Coasting down east side of Mount Hood, above Cooper Spur. Mount Adams in distance.

Mount Hood from the hills south of The Dalles, showing the comparatively timberless country east of the Cascades. Compare this treeless region, as well as the profile of Mount Hood here shown, with the view from Larch Mountain.

Mount Hood, seen from Larch Mountain, on the Columbia River. View looking southeast across the heavily forested ranges of the Cascades to the deep canyons below Ladd and Sandy glaciers.

ley toward Mount Hood, or northward, up the White Salmon Valley toward Trout Lake and Mount Adams. Here is unrolled a wealth of fertile lowlands, surrounded by lofty ranges made beautiful by their deep forests and rising to grandeur in their snow-peaks.

Butterfly on the summit of Mount Hood.

Leaving the canyon of the Columbia, in either direction the road follows swift torrents of white glacial water that tell of a source far above. It crosses a famous valley, among its orchards and hayfields, but always in view of the dark blue mountains and of the snow-covered volcanoes that rise before and behind, their glaciers shining like polished steel in the sunlight. So the visitor reaches the foot of his mountain. Losing sight of it for a time, he follows long avenues of stately trees as he climbs the benches. In a few hours he stands upon a barren shoulder of the peak, at timber line. A new world confronts him. The glaciers reach their icy arms to him from the summit, and he breathes the winds that sweep down from their fields of perennial snow.

Members of Portland Snow-shoe Club on way to Mount Hood in Winter, and at their club house, near Cloud Cap Inn.

It is all very different from Switzerland, this quick ascent from bending orchards and forested hills to a mighty peak standing white and beautiful in its loneliness. But it is so wonderful that Americans who love the heights can no longer neglect it, and each year increasing numbers are discovering that here in the Northwest is mountain scenery worth traveling far to see, with very noble moun-

Fumarole, or gas vent, near Crater Rock.

Looking across the head of Eliot glacier from near the summit of Mount Hood.

tains to climb, true glaciers to explore, and the widest views of grandeur and interest to enjoy. Such sport combines recreation and inspiration.

The traveler from Portland to either Mount Hood or Mount Adams may go by rail or steamer to Hood River, Oregon, or White Salmon, Washington. These towns are on opposite banks of the Columbia at its point of greatest beauty. Thence he will journey by automobile or stage up the corresponding valley to the snow-peak at its head. If he is bound for Mount Hood his thirty-mile ride will bring him to a charming mountain hotel, Cloud Cap Inn, placed six thousand feet above the sea, on a ridge overlooking Eliot glacier, Hood's finest ice stream.

Mount Hood at night, seen from Cloud Cap Inn. This view is from a negative exposed from nine o'clock until midnight.

If Mount Adams be his destination, a ride of similar length from White Salmon will bring him merely to the foot of the mountain. The stages run only to Guler, on Trout Lake, and to Glenwood. Each of these villages has a comfortable country hotel which may be made the base for fishing and hunting in the neighborhood. Each is about twelve miles from the snow-line. At either place, guides, horses and supplies may be had for the trip to the mountain. Glenwood is nearer to the famous Hellroaring Canyon and the glaciers of the southeast side. Guler is a favorite point of departure for the south slope and for the usual route to the summit.

Another popular starting point for Mount Adams is Goldendale, reached by a branch of the North Bank railway from Lyle on the Columbia. This route also leads to the fine park district on the southeastern slope, and it has a special attraction, as it skirts the remarkable canyon of the Klickitat River. Many parties also journey to the mountain from North Yakima and other towns on the Northern Pacific railway. Hitherto, all such travel from either north or south has meant a trip on foot or horseback over interesting mountain trails, and has involved the necessity of packing in camp equipment and supplies. During the present summer, a hotel is to be erected a short distance from the end of Mazama glacier, at an altitude of about sixty-five hundred feet, overlooking Hellroaring Canyon on one side, and on the other a delightful region of mountain tarns, waterfalls and alpine flower meadows. Its verandas will command the Mazama and Klickitat glaciers, and an easy route will lead to the summit. With practicable roads from Goldendale and Glenwood, it should draw hosts of lovers of scenery and climbing, and aid in making this great mountain as well known as it deserves to be.

Climbing Mount Hood, with ropes anchored on the summit and extending down on east and south faces of the peak.

Visitors going to Mount Hood from Portland have choice of a second very attractive hotel base in Government Camp, on the south slope at an altitude of thirty-nine hundred feet. This is reached by automobiles from the city,

North side of Mount Hood, seen from moraine of Coe glacier. This glacier flows down from the summit, where its snow-field adjoins that of Eliot glacier (left). West of the Coe, the Ladd glacier is seen, separated from the former by Pulpit Rock, the big crag in the middle distance, and Barrett Spur, the high ridge on the right.

COPYRIGHT, G. M. WEISTER

Looking West on summit of Mount Hood, with Mazama Rock below.

over a fair road that will soon be a good road, thanks to the Portland Automobile Club. The mountain portion of this highway is the historic Barlow road, opened in 1845, the first wagon road constructed across the Cascades. As the motor climbs out of the Sandy River valley, and grapples the steep moraines built by ancient icefields, the traveler gets a very feeling reminder of the pluck of Captain Barlow and his company of Oregon "immigrants" in forcing a way across these rugged heights. But the beauty of the trip makes it well worth while, and Government Camp gives access to a side of the peak that should be visited by all who would know how the sun can shatter a big mountain with his mighty tools of ice.

The hotel here was erected in 1900 by O. C. Yocum, under whose competent guidance many hundreds of climbers reached the summit of Mount Hood. The Hotel is now owned by Elisha Coalman, who has also succeeded to

Summit of Mount Hood, from Mazama Rock, showing the sun-cupped ice of midsummer.

Mount Hood, seen from Sandy River canyon, six miles west of snow line. This important picture begins with Barrett Spur and Ladd glacier on the north sky line (left). On the northwest face of the peak is the main Sandy glacier, its end divided by a ridge into two parts. The forested "plowshare" projecting into the canyon is Yocum Ridge. South of it the south branch of the Sandy river flows down from a smaller glacier called the Little Sandy, or Reid. The broad bottom of this canyon and the scored cliffs on its sides show that it was formerly occupied by the glacier.

his predecessor's office as guide. During the last year he has enlarged his inn, and he is now also building comfortable quarters for climbers at a camp four miles nearer the snow line, on the ridge separating White River glacier from Zigzag glacier.

MOUNT HOOD.

Mount Hood is the highest mountain in Oregon, and because of a general symmetry in its pyramidal shape and its clear-cut, far-seen features of rock and glacier, it has long been recognized as one of the most beautiful of all American snow peaks. Rising from the crest of the Cascades, it presents its different profiles and variously sculptured faces to the entire valley of the Columbia, east and west, above which it towers in stately magnificence, a very king of the mountains, ruling over a domain of ranges, valleys and cities proud of their allegiance.

Crevasses on Coe glacier.

On October 20, 1792, Lieutenant Broughton, of Vancouver's exploring expedition in quest of new territories for His Majesty George III., discovered from the Columbia near the mouth of the Willamette, "a very distant high snowy mountain, rising beautifully conspicuous," which he strangely mistook to be the source of the great river. Forthwith he named it in honor of Rear Admiral Samuel Hood, of the British Admiralty who had distinguished himself in divers naval battles during the American and French Revolutions.

The mountain has been climbed more often than any other American snow-peak. The first ascent was made on August 4, 1854, from the south side, by a party under Captain Barlow, builder of the "immigrant road." One of the climbers, Editor Dryer of *The Oregonian*, published an account

Mount Hood, with Crevasses of Eliot Glacier in foreground.

"Evermore the wind
Is thy august companion; yea, thy peers

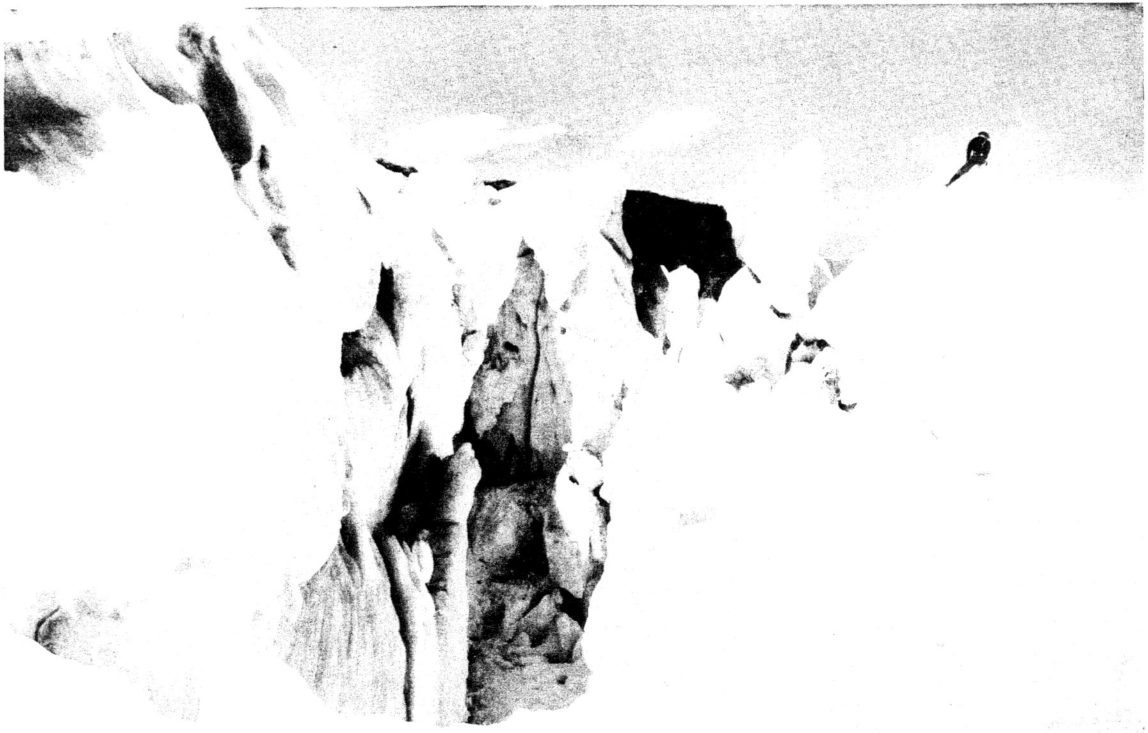

Crevasses and Ice Pinnacles on Eliot Glacier, Mount Hood.

Mount Hood, seen from the top of Barrett Spur. On the left, cascading down from the summit, is Coe glacier; on the right, Ladd glacier. The high cliff separating them is "Pulpit Rock."

of the trip in which, with more exactness than accuracy, he placed the height of the mountain at 18,361 feet! The most notable ascent by a large party took place forty years later, when nearly two hundred men and women met on the summit, and there, with parliamentary dispatch bred of a bitter wind, organized a mountain club which has since become famous. For its title they took the name "mazama," Mexican for the mountain goat, close kin to the Alpine chamois. Membership was opened to those who have scaled a snow-peak on foot. By their publications and their annual climbs, the Mazamas have done more than any other agency to promote interest in our Northwestern mountains.

Mount Hood stands, as I have said, upon the summit of the Cascades. The broad and comparatively level back of the range is here about four thousand feet above the sea. Upon this plane the volcano erected its cone, chiefly by the expulsion of scoriæ rather than by extensive lava flows, to a farther height of nearly a mile and a half. There is no reason to suppose that it ever greatly exceeded its present altitude, which gov-

Ice Cascade, south side of Mount Hood, near head of White River glacier.

Little Sandy or Reid glacier, West side of Mount Hood.

ernment observations have fixed at 11,225 feet. Its diameter at its base is approximately seven miles from east to west.

Compared with Mount Adams, its broken and decapitated northern neighbor, Mount Hood, although probably dating from Miocene time, is still young enough to have retained in a remarkable degree the general shape of its original cone. But as we approach it from any direction, we find abundant proof that powerful destructive agents have been busy during the later geological ages. Already the summit plateau upon which the peak was built up has been largely dissected by the glaciers and their streams. The whole neighborhood of the mountain is a vastly rugged district of glacial canyons and eroded water channels, trenched deep in the soft volcanic ashes and the underlying ancient rock of the range. The mountain itself, although still a pyramid, also has its story of age and loss. Its eight glaciers have cut away much of its mass. On three sides they have burrowed so deeply into the cone that its original angle, which surviving ridges show to have been about thirty degrees, has on the upper glacial slopes been doubled. This is well illustrated by the views shown on pages 58, 61, 69 and 71.

Portland Y. M. C. A. party starting for the summit at daybreak. South side of Mount Hood.

This cutting back into the mountain has greatly lessened the area of the upper snow-fields. The reservoirs feeding the glaciers, are therefore much smaller than of old, but, by

Crater of Mount Hood, seen from south side. Its north rim is the distant summit ridge. Steel's Cliff (right) and Illumination Rock (left) are parts of east and west rims. The south wall has been torn away, but the hard lava core remains in Crater Rock, the cone rising in center. Note the climbers ascending the "Hog-back" or ridge leading from Crater Rock up to the "bergschrund," a great crevasse which stretches across the crater at head of the glaciers. The ridge in foreground is Trian[illegible]le Moraine. On its ri[illegible]

South side of Mount Hood, seen from crag on Tom-Dick-and-Harry Ridge, five miles from the snow-line. A thousand feet below is the hotel called "Government Camp," with the Barlow road, the first across the Cascades. On left are Zigzag and Sand canyons, cut by streams from Zigzag glacier above.

way of compensation, present a series of most interesting ice formations on the steeper slopes. In this respect, Mount Hood is especially noteworthy among our Northwestern snow-peaks. While larger glaciers are found on other mountains, none are more typical. The glaciers of Hood especially repay study because of their wonderful variety of ice-falls, terraces, seracs, towers, castles, pinnacles and crevasses. Winter has fashioned a colossal architecture of wild forms.

Ye ice-falls! ye that from the mountain's brow
Adown enormous ravines slope amain,—
Torrents, methinks, that heard a mighty voice,
And stopped at once amid their maddest plunge!
Motionless torrents! silent cataracts!

Crag on which above view was taken.

The visitor who begins his acquaintance with Mount Hood on the north side has, from Cloud Cap Inn, four interesting glaciers within a radius of a few miles. Immediately before the Inn, Eliot glacier displays its entire length of two miles, its snout being only a few rods away. West of this, Coe and Ladd glaciers divide the north face with the Eliot. All three have their source in neighboring reservoirs near the summit, which have been greatly reduced in area. This, with the resulting shrinkage in the glaciers, is shown by the high lateral moraines left as the width of the ice streams has lessened. On the east slope is a fine cliff glacier, the Newton Clark, separated from the Eliot by Cooper Spur, a long ridge that furnishes the only feasible north-side route for climbers to the summit.

Part of the "bergschrund" above Crater Rock. A bergschrund is a crevasse of which the lower side lies much below its upper side. It is caused by a sharp fall in the slope, or by the ice at the head of a glacier pulling away from the packed snow above.

Climbing Cooper Spur is a tedious struggle up a long cinder slope, but it has its reward in fine views of the near-by glaciers and a wide outlook over the surrounding country. A tramp of three miles from the Inn covers the easier grade, and brings the climber to a height of eight thousand feet. A narrow, snow-covered chine now offers a windy path to the foot of the steeper slope (See p. 60). The climb ends with the conquest of a half-mile of vertical elevation over a grade that tests muscle, wind and nerve. This is real mountaineering, and as the novice clutches the rocks, or carefully follows in the steps cut by the guide, he recalls a command well adapted to such trying situations: "Prove all things; hold fast that which is good." But the danger is more apparent than real, and the goal is soon reached.

Prof. Harry Fielding Reid and party exploring Zigzag glacier, south side of Mount Hood. Illumination Rock is seen beyond.

The south-side route, followed by the Barlow party of 1854, was long deemed the only practicable trail to the summit. Many years later, William A. Langille discovered the route up from Cooper Spur. The only accident charged against this path befell a stranger who was killed in trying to climb it without a guide. Its steepness is, indeed, an advantage, as it requires less time than the other route. Climbers frequently ascend by one trail and descend by the other, thus making the trip between Cloud Cap Inn and Government Camp in a day.

The actual summit of Mount Hood is a narrow but fairly level platform, a quarter of a mile long, which is quickly seen to be part of the rim of the ancient crater. Below it, on the north, are the heads of three glaciers already mentioned, the Eliot, Coe and Ladd; and looking down upon them, the climber perceives that here the mountain has been so much cut away as to be less a slope than a series of precipices, with very limited benches which serve as gathering grounds of snow. (See pp. 55, 67 and 70.) These shelves feed the lower ice-streams with a diet of avalanches that is year by year becoming less bountiful as this front becomes more steep. Soon, indeed, geologically speaking, the present summit, undermined by the ice, must fall, and the mountain take on a new aspect, with a lower, broader top. Thus while the beautiful verse which I have quoted under the view of Mount Hood from White Salmon (p. 56) is admirable poetry, its last line is very poor geology. This, however, need not deter any present-day climbers!

Mazamas climbing the "Hog-back," above Crater Rock, and passing this rock on the descent.

On the south side of the summit ridge a vastly different scene is presented. Looking down over its easy slope, one recognizes even more clearly than from the north-side view that Mount Hood is merely a wreck of its former graceful cone, a torn and disintegrating remnant, with very modest pretentions to symmetry, after all, but still a fascinating exhibit of the work of such Gargantuan forces as hew and whittle such peaks.

Portland Ski Club on south side of Mount Hood, above Government Camp.

The crater had a diameter of about half a mile. Its north rim remains in the ridge on which our climber stands. All the rest of its circumference has been torn away, but huge fragments of its wall are seen far below, on the right and left, in "cleavers" named respectively Illumination Rock and Steel's Cliff. One of these recalls several displays of red fire on the mountain by the Mazamas. The other great abutment was christened in honor of the first president of that organization.

Apart from these ridges, the entire rim is missing; but below the spectator, at what must have been the center of its circle, towers a great cone of lava, harder than the andesitic rocks and the scoriae which compose the bulk of the mountain. This is known as Crater Rock. It is the core of the crater, formed when the molten lava filling its neck cooled and hardened. Around it the softer mass has worn down to the general grade of the south slope, which extends five miles from just below the remaining north rim at the head of the glaciers to the neighborhood of Government Camp, far down on the Cascade plateau. The grade is much less than thirty degrees. Over the slope flow down two glaciers, the Zigzag on the west, and the White River glacier on the east, of Crater Rock.

Mount Hood Lily.
(*L. Washingtonianum*)

It is sometimes said that the south side of the old summit was blown away by a terrific explosion. That is improbable, in view of Crater Rock, which indicates a dormant volcano when the south side was destroyed. The mountain was doubtless rent by ice rather than by fire. The mass of ice and snow in and upon the crater broke apart the comparatively loose wall, and pushed its shattered tuffs and cinders far down the slopes. Forests were buried, old canyons were filled, and the whole southwest side of the mountain was covered with the fan-shaped outwash from the breach. Through this debris of the ancient crater the streams at the feet of the glaciers below are cutting vast ravines which can be seen from the heights above. (See illustrations, pp. 77-81.)

The central situation of Mount Hood makes the view from its summit especially worth seeking. From the Pacific to the Blue Mountains, south almost to the California line, and north as far, it embraces

Mazama party exploring White River glacier, Mount Hood.

an area equal to a great state, with four hundred miles of the undulating Cascade summits and a dozen calm and radiant snow-peaks. The Columbia winds almost at its foot, and a multitude of lakes, dammed by glacial moraines and lava dikes, nestle in its shadow. This view "covers more history," as Lyman points out, than that from any other of our peaks. About its base the Indians hunted, fished and warred. Across its flank rolled the great tide of Oregon immigration, in the days of the ox-team and settler's wagon. It has seen the building of two states. It now looks benignly down upon the prosperous agriculture and growing cities of the modern Columbia basin, and no doubt contemplates with serenity the time when its empire shall

NeWton Clark glacier, east side of Mt. Hood, seen from Cooper Spur, With Mt. Jefferson fifty miles south.

be one of the most populous as it is one of the most beautiful and fertile regions in America. No wonder the shapely mountain lifts its head with pride!

Returning to the glaciers of the north side, we note that all three end at an altitude close to six thousand feet. None of them has cut a deep, broad bed for itself like the great radiating canyons which dissect the Rainier National Park and protect its glaciers down to a level averaging four thousand feet. Instead, these glaciers lie up on the side of Mount Hood, in shallow beds which they no longer fill; and are banked between double and even tripple border moraines, showing successive advances and retreats of the glaciers. (See illustration, top of p. 59.) The larger moraines stand fifty to a hundred feet above the present ice-streams, thus indicating the former glacier levels. No vegetation appears on these desolate rock and gravel dikes. The retreat of the glaciers was therefore comparatively recent.

Looking from Mount Jefferson, along the summits of the Cascades, to Mount Hood.

COPYRIGHT, G. M. WEISTER

Shadow of Mount Hood, seen from Newton Clark glacier shortly before sunset. View shows two branches of East Fork of Hood River, fed by the glacier, and the canyon of the East Fork, turning north. Beyond it (left) are Tygh Hills and wheat fields of the Dufur country. On the right is Juniper Flat, with the Deschutes canyon far beyond.

Eliot glacier has been found by measurement near its end, to have a movement of about fifty feet a year. On the steeper slope above, it is doubtless much greater. All the three glaciers are heavily covered, for their last half mile, with rocks and dirt which they have freighted down from the cliffs above, or dug up from their own beds in transit. None of the lateral moraines extends more than two or three hundred yards below the snout of its glacier. Each glacier, at its end, drops its remnant of ice into a deep V-shaped ravine, in which, not far below, trees of good size are grow-

Snout of Newton Clark glacier.

Mount Hood and Hood River, seen from a point twenty miles north of the mountain.

Lava Flume near Trout Lake, about thirty feet wide and forty feet high.

ing. Hence it would not seem that these north-side glaciers have ever extended much farther than they do at present. The ravine below Eliot glacier, however, half a mile from the snout, is said to show glacial markings on its rocky sides. It is evident, in any case, that the deep V cuttings now found below the glaciers are work of the streams. If these glaciers extended farther, it was at higher levels than their present stream channels. As the glaciers receded, their streams have cut the deep gorges in the soft conglomerates. Between Eliot and Coe glaciers are large snow-fields, ending much farther up than do the glaciers; and below these, too, the streams have trenched the slope (See illustration, p. 57.)

Between Coe and Ladd glaciers is a high rocky ridge known as Barrett Spur, from which, at nearly 8,000 feet, one may obtain glorious views of the peak above, the two glaciers sweeping down its steep face and the sea of ranges stretching westward. (See illustrations, pp. 69 and 75.) Barrett Spur may have been part of the original surface of the mountain, but is more likely the remnant of a secondary cone, ice and weathering having destroyed its conical shape. From its top, the climber looks over into the broad-bottomed canyon of Sandy River, fed by the large and small Sandy glaciers of the west slope. (See pp. 71 and 76.) This canyon and that of the Zigzag River, south of it, from Zigzag glacier, are "plainly glacier-sculptured," as Sylvester declares. The same is true of the canyon lying below the White River glacier, on the southeast slope. In journeying to Government Camp, one may

Y. M. C. A. party from North Yakima at Red Butte, an extinct volcano on north side of Mount Adams.

see abundant evidence of the glacial origin of the Sandy and Zigzag canyons. The White River Canyon has been thoroughly explored and described by Prof. Reid.

All three of these wide U-shaped canyons were once occupied by great glaciers, which left their record in the scorings upon the sides of the gorges; in the mesas of finely ground moraine which they spread over the bottoms and through which the modern rivers have cut deep ravines; in trees broken and buried by the glaciers in this drift; in the fossil ice lying beneath it, and in huge angular boulders left standing on the valley floors, several miles from the mountain.

Ice Cave in lava beds near Trout Lake.

Sandy glacier extends three hundred feet farther down the slope than do the north-side glaciers, but the Zigzag and White River glaciers, flowing out of the crater, end a thousand feet higher. This is due not only to the smaller reservoirs which feed them and to their southern exposure, but also doubtless to the easier grade, which holds the ice longer on the slope. On the east side of the peak is a broad ice-stream, the Newton Clark glacier, which also ends at a high altitude, dropping its ice over a cliff into deep ravines at the head of East Fork of Hood River. This glacier, well seen from Cooper Spur, completes the circuit of the mountain. (See pp. 83 and 84.)

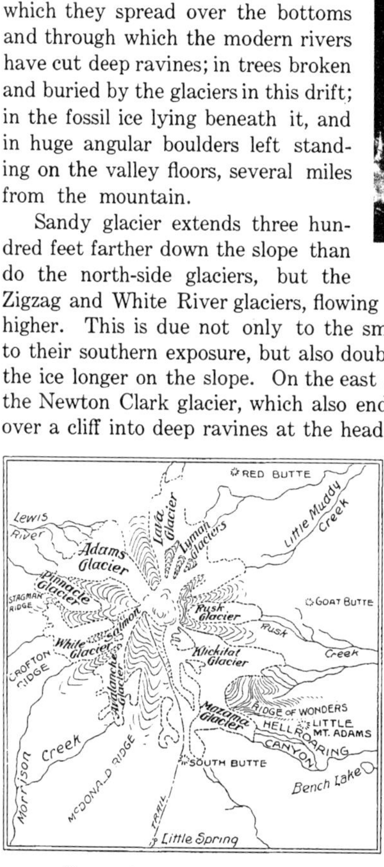

Mount Adams, elevation 12,307 feet.

Sylvester suggests that Mount Hood may not be extinct but sleeping. For this, however, there is little more evidence that may be discovered on other Northwestern peaks. About Crater Rock, steam jets are found, gas escapes, and the rocks are warm in many places. "Fumaroles" exist, where the residuary heat causes openings in the snow bed. Sylvester reports dense smoke and steam issuing from Crater Rock by day and a brilliant illumination there at night, in

Telephoto view of Mount Adams, from the northeast side of Mount St. Helens, at elevation of 7,000 feet, overlooking the densely timbered ranges of the Cascades.

Mount Adams from Trout Creek, at Guler, near Trout Lake; distance twelve miles.

August, 1907. But volcanoes sometimes contradict prophecy, and no further intimations of trouble having since been offered, this display may be deemed the last gasp of a dying monster rather than an awakening toward new life.

MOUNT ADAMS.

Going up the White Salmon Valley toward Mount Adams, the visitor quickly realizes that he is in a different geological district from that around Mount Hood. The Oregon peak is mainly a pile of volcanic rocks and cinders ejected from its crater. Little hard basalt is found, and in all its circumference I know of only one large surface area of new lava. This is a few miles north of

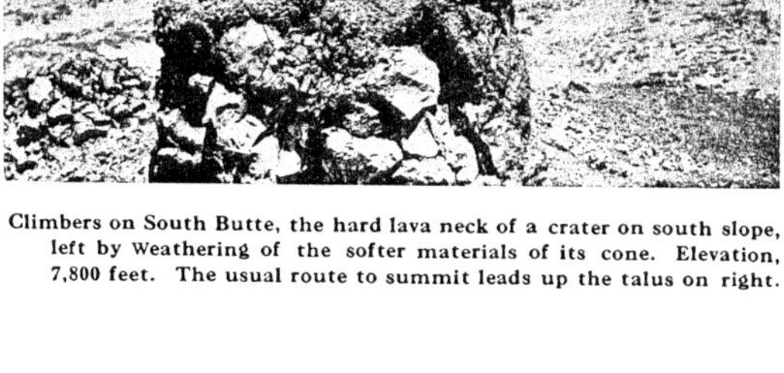

Climbers on South Butte, the hard lava neck of a crater on south slope, left by weathering of the softer materials of its cone. Elevation, 7,800 feet. The usual route to summit leads up the talus on right.

Dawn on Mount Adams, telephotographed from Guler, at 4 a. m., showing the three summit peaks, of which the middle one is the highest. The route of the climbers is up the south slope, seen on right.

Cloud Cap, and so recent that no trees grow on it. But north of the Columbia, one meets evidences of comparatively recent lava sheets in many parts of the valley. Some obviously have no connection with Mount Adams; they flowed out of fissures on the ridges. But these beds of volcanic rock become more apparent, and are less covered with soil, as we approach the mountain, until, long before timber line is reached, dikes and streams of basalt, as yet hardly beginning to disintegrate, are found on all sides of the peak.

The form and slope of Mount Adams tell of an age far greater than Mount Hood's, but its story is not, like that of Hood, the legible record of a simple volcanic cone. It wholly lacks the symmetry of such a pile. Viewed from a distance, it sits very majestically upon the summit of one of the eastern ranges of the Cascades. As we approach, however, it is seen to have little of the conical shape of Hood, still less that of graceful St. Helens, which is young and as yet practically unbroken. Its summit has been much worn down by ice or perhaps by

Foraging in the snow. The Mount Adams country supports hundreds of large flocks of sheep.

Steel's Cliff, southeast side of Mount Hood. In the distance is seen Juniper Flat, in eastern Oregon.

Ice Castle and great Crevasse, near the head of Eliot Glacier, Mt. Hood.

"Touched by a light that hath no name,
A glory never sung,
Aloft on sky and mountain wall

Mazamas climbing a 40° stairway of shattered basalt, north side of Mount Adams.

explosions. Some of its sides are deeply indented, and all are vastly irregular in angle and markings—here a face now too steeply cut to hold a glacier, but showing old glacial scorings far down its slope; there another terraced and ribbed with waves and dikes of lava. The mountain is a long ridge rather than a round peak, and close inspection shows it to be a composite of several great cones, leaning one upon another,—the product of many craters acting in successive ages. On its ancient, scarred slopes, a hundred modern vents have added to the ruggedness and interest of the peak. Many of these blowholes built parasitic cones, from which the snows of later centuries have eroded the loose external mass, leaving only the hard lava cores upstanding like obelisks. Other vents belched out vast sheets of rock that will require a century more of weathering to make hospitable even to the sub-alpine trees most humble in their demands for soil.

Mount Adams from one of the many lakes on its southeast slope. On ridge above, near the end of Mazama glacier, a hotel is to be erected.

Mount Adams therefore presents a greater variety of history, a more complex and fascinating problem for the student to unravel, than any of its neighbors. This interest

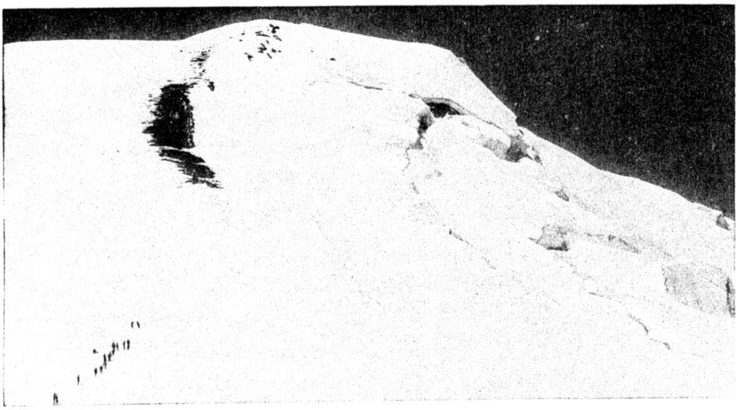

Climbers ascending from South Peak to Middle Peak on Mount Adams, with the "bergschrund" above Klickitat glacier on right. This central dome is about 500 feet higher than South Peak.

extends to the district about it, a country of new lava flows covering much of the older surface. The same conditions mark the region surrounding the newer peak, St. Helens, thirty miles west. In each district, sheets of molten rock have been poured across an ancient and heavily forested land. Thus as we travel up the rich valleys leading from the Columbia to either peak, we meet everywhere the phenomena of vulcanism.

The lava sheet flowing around or over a standing or fallen tree took a perfect impression of its trunk and bark. Thousands of these old tree casts are found near both Adams and St. Helens. Where the lava reached a water-

Mount Adams, seen from Happy Valley, south side. Elevation about 7,000 feet. Mazama glacier is on right.

Mount Adams, from Snow-Plow Mountain, three miles southeast of the snow line; elevation 5,070 feet, overlooking the broad "park" country west of Hellroaring Canyon.

course, it flowed down in a deeper stream,—a river of liquid rock. Lava is a poor conductor of heat; hence the stream cooled more quickly on the surface than below. Soon a crust was formed, like the ice over a creek in winter. Under it the lava flowed on and out, as the flood stopped, leaving a gallery or flume. Later flows filled the great drain again and again, adding new strata to its roof, floor and sides, and lessening its bore. Long after the outflows ceased, weathering by heat and frost broke openings here and there. Many of the flumes were choked with drift. But others, in the newer lava beds, may be explored for miles. It was from the lava caves of northern California that the Modoc Indians waged their famous war in the Seventies.

The disintegration of the lava galleries in the Mount Adams field has of course produced caves of all sorts and sizes. Where one of these is closed at one end with debris, so that the summer air

COPYRIGHT, S. C. SMITH

Wind-Whittled ice near the summit of Mount Adams.

cannot circulate to displace the heavier cold remaining from winter, the cave, if it has a water supply, becomes an ice factory. The Trout Lake district has several interesting examples of such *glacieres*, as they have been named, where one may take refuge from July or August heat above ground, and, forty feet below, in a cave well protected from sun and summer breeze, find great masses of ice, with more perhaps still forming as water filters in from a surface lake or an underground spring. The Columbia River towns as far away as Portland and The Dalles formerly obtained ice from the Trout Lake caves, but at present they supply only some near-by farmers.

Mount Adams is ascended without difficulty by either its north or south slope. On the east and west faces, the cliffs and ice cascades appall even the expert alpinist. As yet, so far as I can learn, no ascents have been made over these slopes. The southern route is the more popular one. It leads by well-marked trails up from Guler or Glenwood, over a succession of terraces clad in fine, open forest; ascends McDonald Ridge, amid increasing barriers of lava; passes South Butte, a decaying pillar of red silhouetted against the black rocks and white snow-fields·,crosses many a caldron of twisted and broken basalt,—"Devil's Half Acres" that once were the hot, vomiting mouths of drains from the

Mazama glacier, at head of Hellroaring Canyon. Upper view shows floor of canyon, a mile below the glacier, with the "Ridge of Wonders" on right. Lower view is from ridge west of the canyon, near end of Mazama glacier, elevation nearly 7,000 feet. Note great lateral moraine which the glacier has built on left.

fiery heart of the peak; scales a giants' stairway tilted to to forty degrees, overlooking the west branch of Mazama glacier on one side and a small unnamed glacier on the other; and at last gains the broad shoulder which projects far on the south slope. (See illustrations, pp. 89 and 93.)

Nearing the summit, south side.

Here, from a height of nine thousand feet, we look down on the low, wide reservoir of Mazama glacier on the east, and up to the ice-falls above Klickitat glacier on the higher slopes beyond. The great platform on which we stand was built up by a crater, three thousand feet below the summit. The climb to it has disclosed the fact that the mountain is composed mostly of lava. Some of the ravine cuttings have shown lapilli and cinders, but these are rarer than on the other Northwestern peaks. The harder structure has resisted the erosion which is cutting so deeply into the lower slopes of Hood. On Mount Adams, not only do the glaciers, with one or two notable exceptions, lie up on the general surface of the mountain, banked by their moraines; but their streams have cut few deep ravines.

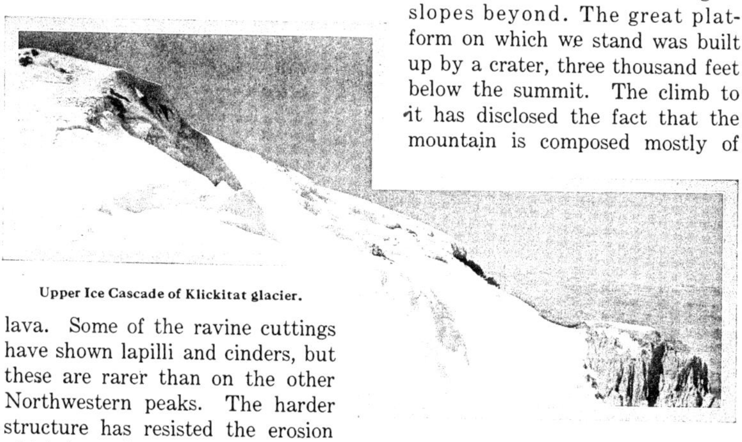

Upper Ice Cascade of Klickitat glacier.

From this point, the route becomes steeper, but is still over talus, until the first of the three summit elevations, known as South Peak, is reached. This is only five hundred feet below the actual summit, Middle Peak, which is gained by a short, hard pull, generally

An Upland "Park," West of Hellroaring Canyon.

Mount Adams, from the Ridge of Wonders, showing the great amphitheater or "cirque" of Klickitat glacier, fed by avalanches from the summit plateau. This is the most important example of glacial sculpture on the mountain. Beyond, on the right, is seen the head of Rusk glacier, while on the left is Mazama glacier. Note the stunted sub-alpine trees scattered thinly over this ridge, even up to an altitude of 7,000 feet.

Storm on Klickitat Glacier, seen from the Ridge of Wonders.

over snow. (See p. 94.) The north-side route is up a long, sharp ridge between Lava and Adams glaciers (p. 104). Like the other path, its grade is at first easy; but its last half mile of elevation is achieved over a slope even steeper, and ending in a longer climb over the snow. Neither route, however, offers so hard a finish as that which ends the Mount Hood climb. From the timber-line on either side, the ascent requires six or seven hours.

The summit ridge is nearly a mile long and two-thirds as wide. It is the gathering ground of the snows that feed Klickitat, Lyman,

Snow cornice above the bergschrund at head of Klickitat glacier, with another part of the same crevasse.

Adams and White Salmon glaciers. (See map, p. 87.) Mazama, Rusk, Lava, Pinnacle and Avalanche glaciers lie beneath cliffs too steep to carry ice-streams. Their income is mainly collected from the slopes, and if they receive snow from the broad summit at all, it is chiefly in the avalanches of early summer. Nearly all the glaciers, however, are thus fed in part, the steep east and west faces making Mount Adams famous for its avalanches.

From the summit on either side, the climber may look down sheer for half a mile to the reservoirs and great ice cascades of the glaciers below. It is

Mount Adams, seen from the northeast, with the Lyman glaciers in center, Rusk glacier on extreme left, and Lava glacier, right. The ridge beyond Lava glacier is the north-side route to the summit. The Lyman glaciers, like Adams glacier on the northwest side, are noteworthy for their cascades of ice.

COPYRIGHT, ASAHEL CURTIS

Mount Adams from Sunnyside, Washington, with irrigation "ditch" in foreground.

seen that with the exception of the Rusk and Klickitat, which are deeply embedded in canyons, the glaciers spread out, fan-like, on the lower slopes, and are held up by their moraines. Most of them end at elevations considerably above six thousand five hundred feet. The difference in this respect between Adams and Hood is due, no doubt, to lighter rainfall.

Crevasse in Lava glacier, north side of Mount Adams.

Of the two glaciers just mentioned the Klickitat is the larger and more typical. The Rusk, however, is of interest because it flows, greatly crevassed, down a narrow flume or couloir on the east slope. Its bed, Reid suggests, may have been the channel of "a former lava flow, which, hardening on the surface, allowed the liquid lava inside to flow out; and later the top broke in." The Klickitat glacier lies in a much larger canyon, which it has evidently cut for itself. This is one of the most characteristic glacial amphitheaters in America, resembling, though on a smaller scale, the vast Carbon glacier *cirque* which is the crowning glory of the Rainier National Park. The Klickitat basin is a mile wide. Into it two steep ice-streams cascade from the summit, and avalanches fall from a cliff which rises two thousand feet between them. (See pp. 98 and 99.)

The glacier is more than two miles long. It ends at an elevation of less than six thousand feet, covered with debris from a large medial moraine formed by the junction of the two tributary glaciers. Like the other Mount Adams glaciers, and indeed nearly all glaciers in the northern hemisphere, it is shrinking, and has built several moraines on each side. These extend half a mile below its present snout, and the inner moraines are underlaid with ice, showing the retreat has been recent.

North Peak of Mount Adams, with The Mountaineers beginning their ascent, in 1911. Their route led up the ridge seen here, which divides Lava glacier, on the left, from Adams glacier, on extreme right.

South of the Klickitat glacier, a part of the original surface of the peak remains in the great Ridge of Wonders. Rising a thousand feet above the floor of Hellroaring Canyon, which was formerly occupied by Mazama glacier, now withdrawn to the slope above, this is the finest observation point on the mountain. "The wonderful views of the eastern precipices and glaciers," says Reid, "the numerous dikes, the well preserved parasitic cone of Little Mount Adams, and the curious forms of volcanic bombs scattered over its surface entirely justify the name Mr. Rusk has given to this ridge."

Snow Bridge over Killing Creek, north of Mount Adams.

Adams glacier, upon the northwest slope, with a length of three miles, is the largest on the mountain. This and the two beautiful ice streams on the northeast, named after Prof. W. D. Lyman, are notable for their ice-falls, half-mile drops of tumbling, frozen rivers.

The naming of the mountain was a result of the movement started by Hall J. Kelley, the Oregon enthusiast, in 1839. The northwestern snow-peaks, so far as shown in maps of the period, bore the names given by

North-side Cleaver, with Lava glacier on left. This sharp spine was climbed by The Mountaineers and the North Yakima Y. M. C. A. party in 1911.

Vancouver as part of his annexation for George III. The utility, beauty and historic fitness of the significant Indian place names did not occur to a generation busy in ousting the Indian from his land; but our grandfathers remembered George III. Kelley and other patriotic men of the time proposed to call the Cascades the "Presidents' Range," and to christen the several snow-peaks for individual ex-presidents of the United States. But the second quarter of the last century knew little about Oregon, and cared less. The well-meant but premature effort failed, and the only names of the presidents which have stuck are Adams and Jefferson. Lewis and Clark mistook Mount Adams for St. Helens, and estimated it "perhaps the highest pinnacle in America." The Geological Survey has found its height to be 12,307 feet. Mount Adams was first climbed in 1854 by a party in which were Col. B. F. Shaw, Glenn Aiken and Edward J. Allen.

MOUNT ST. HELENS.

The world was indebted for its first knowledge of Mount St. Helens to Vancouver. Its name is one of the batch which he fastened in 1792 upon our Northwestern landmarks. These honored a variety of persons, ranging from Lord St. Helens, the diplomat, and pudgy Peter Rainier, of the British Admiralty, down to members of the explorer's crew.

Looking across Adams glacier, northwest side of Mount Adams, from ridge shown above.

The youngest of the Cascade snow-peaks, St. Helens is also the most symmetrical in its form, and to many of its admirers the most beautiful. Unlike Hood and Adams, it does not stand upon the narrow summit of one of

"The Mountain that Was 'God'," the great peak which the Indians reverenced and named "Tacoma," seen above the clouds of a rainy day, from the summit of Mount Adams, distant forty miles.

is," said a well-known lecturer, as the
ure was thrown upon his screen, "is the
e the angels look down upon!"

the Cascade ranges, but rises west of the main ridges of that system from valley levels about one thousand feet above the sea. Surrounded by comparatively low ridges, it thus presents its perfect and impressive cone for almost its entire height of ten thousand feet.

The mountain is set well back from the main traveled roads, in the great forest of southwestern Washington. It is the center of a fine lake and river district which attracts sportsmen as well as mountain climbers. A large company visiting it must carry in supplies and camp equipment, but small parties may find accommodation at Spirit Lake on the north, and Peterson's rȧnch on Lewis River, south of the peak. The first is four, the second is eight, miles

Northwest slope of Mount Adams, with Adams glacier, three miles long, the largest on the mountain. It has an ice-fall of two thousand feet. The low-lying reservoir of Pinnacle glacier is on extreme right, and the head of Lava glacier on left.

from the snow line. Visitors from Portland, Tacoma or Seattle, bound for the north side, leave the railway at Castle Rock, whence a good automobile road (forty-eight miles) leads to the south side of Spirit Lake. Peterson's may be reached by road from Woodland (forty-five miles) or from Yacolt (thirty miles). Well-marked trails lead from either base to camping grounds at timber line. The mountain is climbed by a long, easy slope on the south, or by a much steeper path on the north.

Like Mount Adams, St. Helens is largely built of lava, but the outflows have been more recent here than upon or near the greater peak. The volcano was in eruption several times between 1830 and 1845. The sky at Vancouver

Mount Adams from the southwest, with White Salmon glacier (left) and Avalanche glacier (right) flowing from a common source, the cleft between North and Middle Peaks. The latter, however, derives most of its support from slopes farther to right. Note the huge terminal moraines built by these glaciers in their retreat. Pinnacle glacier is on extreme left.

was often darkened, and ashes were carried as far as The Dalles. To these disturbances, probably, are due the great outflows of new lava covering the south and west sides of the mountain, and much of the country between it and the North Fork of Lewis River. The molten stream flowed westward to Goat Mountain and the "Buttes," of which it made islands; threw a dike across a watercourse and created Lake Merrill; and turning southward, filled valleys and overwhelmed good forest with sheets of basalt. Upon the slope just north of Peterson's, a great synclinal thus buried presents one of the latest pages in the volcanic history of the Columbia basin.

Mount St. Helens, elevation 10,000 feet.

Many hours may be spent with interest upon this lava bed. It is an area of the wildest violence, cast in stone. Swift, ropy streams, cascades, whirling eddies, all have been caught in their course. "Devil's Punch Bowl," "Hell's Kitchen," "Satan's Stairway" are suggestive phrases of local description. The underground galleries here are well worth visiting. Tree tunnels and wells abound. Most important of all, the struggle seen everywhere of the forest to gain a foothold on this iron surface illustrates Nature's method of hiding so vast and terrible a callus upon her face. It is evident that the healing of the wound began as soon as the lava cooled, and that, while still incomplete, it is unceasingly prosecuted. (See p. 111.)

The first volcanic dust from the uneasy crater of St. Helens had no sooner lodged in some cleft opened by the contraction of cooling than a spore or seed carried by the wind or

Scenes in the canyon of the North Fork of Lewis River, fed by the glaciers of Mount Adams and Mount St. Helens.

dropped by a bird made a start toward vegetation. Failing moisture, and checked by lack of soil, the lichen or grass or tiny shrub quickly yielded its feeble existence in preparation for its successor. The procession of rain and sun encouraged other futile efforts to find rootage. Each of these growths

COPYRIGHT, B. A. GIFFORD

Columbia River and Mount Adams, seen from Hood River, Oregon.

"And forests ranged like armies, round and round
At feet of mountains of eternal snow;
And valleys all alive with happy sound,—
The song of birds; swift streams' delicious flow;
The mystic hum of million things that grow."— Helen Hunt Jackson

Southwest side of Mount Adams, reflected in Trout Lake, twelve miles south of the mountain.

lengthened by its decay the life of the next. With winter came frost, scaling flakes from the hard surface, or penetrating the joints and opening fissures in the basalt. Further refuge was thus made ready for the dust and seeds and moisture of another season. The moss and plants were promoters

Scenes on great lava field south of Mount St. Helens. The lodgepole pine thicket above shows struggle of forest to gain a foothold on the rich soil slowly forming over new volcanic rock. The peak itself, with stunted forest at its base, is seen next; and below, one of many "tree tunnels," formed when the lava flowed over or around a tree, taking a perfect cast of its bark.

as well as beneficiaries of this disintegration. Their smallest rootlets found the water in the heart of the rocks, and growing strong upon it, shattered their benefactors.

Soon more ambitious enterprises were undertaken. Huckleberry bushes, fearless even of so unfriendly a surface, started from every

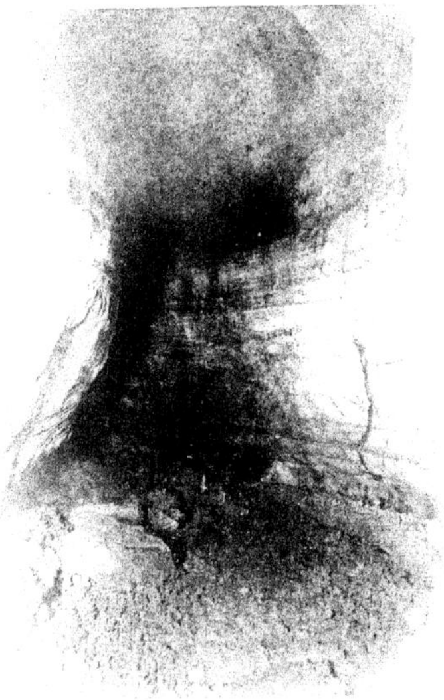

Lava Flume south of Mount St. Helens, a tunnel several miles in length, about twenty feet high and fifteen feet wide.

depression among the rocks. The first small trees appeared. Weakling pines, dwarf firs and alders, shot up for a few feet of hurried growth in the spring moisture, taking the unlikely chance of surviving the later drought. Here and there a seedling outlasted the long, dry summer, and began to be a real tree. Quickly exhausting its little handful of new earth, the daring upstart must have perished had not the melting snows brought help. They filled the hollows with wash from the higher slopes. The treelets found that their day had come, and seizing upon these rich but shallow soil beds, soon covered them with thickets of spindling lodgepole pines and deciduous brush. Such pygmy forests are at length common upon this great field of torn and decaying rock, and all are making their contributions of humus year by year to the support of future tree giants. These will rise by survival of the fittest as the forest floor deepens and spreads.

St. Helens, although much visited, has not yet been officially surveyed or mapped. Its glaciers are not named, nor has the number of true ice-streams been determined. Those on the south and southwest are insignificant. Elsewhere, the glaciers are short and broad, and with one exception, occupy shallow beds. On the southeast, there is a remarkable cleft, shown on page 115, which is doubtless due to volcanic causes rather than erosion, and from which the largest glacier issues. Another typical glacier, distinguished by the finest crevasses and ice-falls on the peak, tumbles down a steep, shallow depression on the north slope, west of the battered parasitic cone of "Black Butte." West of this glacier, in turn, ridges known as the "Lizard" and the "Boot" mark the customary north-side path to the summit. (See p. 118.)

Entrance to Lava Cave shown above. Note strata in roof, showing successive lava flows; also ferns growing from roof.

Telephotograph of Mount St. Helens, from the lower part of Portland, with the summit peaks of Mount Rainier-Tacoma in distance on left, and the Willamette River in foreground.

Mount St. Helens, from Chelatchie Prairie on Lewis River, distance twenty miles. Shows a typical farm clearing in the forest.

Beyond these landmarks, on the west side of the peak, a third considerable glacier feeds South Toutle River. The ravines cut by this stream will repay a visit. (See p. 116.)

The slopes not covered with new lava sheets and dikes exhibit, below the snow-line, countless bombs hurled up from the crater, with great fields of pumice embedding huge angular rocks that tell a story not written on our other peaks. These hard boulders, curiously different from the soft materials in which they lie, were fragments of the tertiary platform on which the cone was erected. Torn off by the volcano, as it enlarged its bore, they were shot out without melting or change in substance. On every hand is proof

Mount St. Helens, seen from Twin Buttes, twenty miles away, across the Cascades. View shows the remarkable cleft or canyon on the southeast face of the peak.

that this now peaceful snow-mountain, which resembles nothing else so much as a well-filled saucer of ice cream, had a hot temper in its youth, and has passed some bad days even since the coming of the white man.

The mountain was first climbed in August, 1853, by a party which included the same T. J. Dryer who, a year later, took part in the first ascent of Mount Hood. In a letter to *The Oregonian* he said the party consisted of "Messrs. Wilson, Smith, Drew and myself." They ascended the south side. The other slopes were long thought too steep to climb, but in 1893 Fred G. Plummer, of Tacoma, now Geographer of the United States Forest Service, ascended the north side. His party included Leschi, a Klickitat Indian, probably the first of his superstitious race to scale a snow-peak. The climbers found

Canyons of South Toutle River, West side of St. Helens. These vast trenches in the soft pumice show by their V shape that they have been cut by streams from the glaciers above, rather than by the glaciers themselves, which, on this young peak, have probably never had a much greater extension.

evidence of recent activity in two craters on the north slope, and photographed a curious "diagonal moraine," as regular in shape as a railway embankment, which connected the border moraines of a small glacier. The north side has since seen frequent ascents.

The Mazamas, who had climbed St. Helens from the south in 1898, again ascended it in 1908, climbing by the Lizard and Boot. This outing furnished the most stirring chapter in the annals of American mountaineering.

Lower Toutle Canyon, seen on left above. Note shattered volcanic bomb.

The north-side route proved unexpectedly hard. After an all-day climb, the party reached the summit only at seven o'clock. The descent after nightfall required seven hours. The risk was great. Over the collar of ice

Northeast side of Mount St. Helens, from elevation of 6,000 feet, with Black Butte on the right.

The Mazamas on summit of St. Helens shortly before sunset. The rocks showing above the snow are parts of the rim of the extinct crater. Mount Adams is seen, thirty-five miles away, on the right, while Rainier-Tacoma is forty-five miles north. Photograph taken at 7:15 p. m. The party did not get back to their camp till long after midnight.

North side of St. Helens in winter, seen from Coldwater Ridge, overlooking Spirit Lake. Shows the long ridge called "the Lizard," because of its shape, with "the Boot" above it. On the northeast slope is "Black Butte," probably a secondary crater.

St. Helens, north side, seen from one mile below snow line. Note the slight progress made by the forest upon the scant soil of the pumice ridges; also, how greatly the angle of the sides, as viewed here at the foot of the peak, differs from that shown in Dr. Lauman's fine picture taken on Coldwater Ridge, five miles north. Both show the mountain from the same direction, but the near view gives no true idea of its steepness. Black Butte is on the left.

near the summit, at a grade of more than sixty degrees, the twenty-five men and women slowly crept in steps cut by the leaders, and clutching a single fifty-foot rope. Later came the bombardment of loose rocks, as the party scattered down the slope. I quote from an account by Frank B. Riley, secretary of the club, who was one of the leaders:

> The safety of the entire party was in the keeping of each member. One touch of hysteria, one slip of the foot, one instant's loss of self-control, would have precipitated the line, like a row of bricks, on the long plunge down the ice cliff. Eight times the party stood poised on its scanty foothold while the rope was lowered. When, after an hour and a half, its last member stepped in safety upon the rocks, there yet lay before it five hours of work ere the little red eyes below should widen into welcoming campfires.
>
> Over great ridges, down into vast snowfields, for hours they plunged and slid, while scouts ahead shouted back warning of the crevasses. On, out of the icy clutch of the silent mountain, they plodded. And then, at last, the timber, and the fires and the hot drinks and the warm blankets and the springy hemlock boughs!

Glacier scenes, north side of Mount St. Helens, east of the "Lizard."

Even this was not the most noteworthy adventure of the outing. One evening, while the Mazamas gathered about their campfire at Spirit Lake, a haggard man dragged himself out of the forest, and told of an injured comrade lying helpless on the other side of the peak. The messenger and two companions —Swedish loggers, all three—had crossed the mountain the morning before. After they gained the summit and began the descent, a plunging rock had struck one of the men, breaking his leg. His friends had dragged him down to the first timber, and while one kept watch, the other had encircled the mountain, in search of aid from the Mazamas.

Immediately a relief party of seven strong men, led by C. E. Forsyth of Castle Rock, Washington, started back over the trailless route by which the messenger had come. All night they scaled ridges, climbed into and out of canyons, waded icy streams. Before dawn they reached the wounded laborer. Mr. Riley says:

It was impossible to carry the man back through the wild country around the peak. Below, the first cabin on the Lewis River lay beyond a moat of forbidding canyons. Above slanted the smooth slopes of St. Helens. Placing the injured man upon a litter of canvas and alpine stocks, they began the ascent of the mountain with their burden. The day dawned and grew old, and still these men crawled upward in frightful, body-breaking struggle. Twelve hours passed, and they had no food and no sleep, save as they fell unconscious downward in the snow, as they did many times, from fatigue and lack of nourishment. At four o'clock, Anderson was again on the summit. Then, without rest, came the descent to the north. Down precipitous cliffs of ice they lowered him, as tenderly as might be; down snow-slopes seared with crevasses, shielding him from the falling rocks; over ridges of ragged lava, until in the deepening darkness of the second night they found themselves again at timber. But in the net-work of canyons they had selected the wrong one, and were lost. Here, at three o'clock, they were found by a second relief party, and guided over a painful five-mile journey home.

Finest of the St. Helens glaciers, north side, with Black Butte on left. It is proposed to call this "Forsyth glacier," in honor of C. E. Forsyth, leader in a memorable rescue.

It was day when camp was reached. In an improvised hospital, a young surgeon, aided by a trained nurse, both Mazamas, quickly set the broken bones. Then they sent their patient comfortably away to the railroad and a Portland hospital. Before the wagon started, Anderson, who had uttered no groan in his two days of agony, struggled to a sitting posture, and searched the faces of all in the crowd about him.

"Ay don't want ever to forget how you look," he said simply; "you who have done all this yust for me."

It is fitting that such an event should be commemorated. With the approval of Mr. Riley and other Mazamas who were present at the time, I would propose that the north-side glacier already described, the most beautiful of the St. Helens ice-streams, be named "Forsyth glacier," in honor of the leader of this heroic rescue.

Road among the Douglas Firs.

Ships loading lumber at one of Portland's large mills.

III.

THE FORESTS

By HAROLD DOUGLAS LANGILLE

As the lowlander cannot be said to have truly seen the element of water at all, so even in his richest parks and avenues he cannot be said to have truly seen trees. For the resources of trees are not developed until they have difficulty to contend with; neither their tenderness of brotherly love and harmony, till they are forced to choose their ways of life where there is contracted room. The various action of trees, rooting themselves in inhospitable rocks, stooping to look into ravines, hiding from the search of glacial winds, reaching forth to the rays of rare sunshine, crowding down together to drink at sweetest streams, climbing hand in hand the difficult slopes, gliding in grave procession over the heavenward ridges—nothing of this can be conceived among the unvexed and unvaried felicities of the lowland forest.

—*Ruskin: "Modern Painters."*

STAND upon the icy summit of any one of the Columbia's snow-peaks, and look north or west or south across the expanse of blue-green mountains and valleys reaching to the sea; your eyes will rest upon the greatest forest the temperate zone has produced within the knowledge of man. Save where axe and fire have turned woodland into field or ghostly "burn," the mantle is spread. Along the broad crests of the Cascades, down the long spurs that lead to the valleys, and across the Coast Range, lies a wealth of timber equaled in no other region. The outposts of this

Outposts of the Forest. Storm-sWept White-bark Pines on Mount Hood.

Alpine Hemlocks at the timber-line on Mt. Adams. Mt. Hood in distance.

great army of trees will meet you far below.

Rimming about your peak, braving winds and the snows that drift in the lee of old moraines, and struggling to break through the timber-line, six thousand feet above the sea, somber mountain hemlocks (*Tsuga mertensiana*) and lighter white-bark pines (*Pinus albicaulis*) form the thin vanguard of the forest. They meet the glaciers. They border the snow-fields. They hide beneath their stunted, twisted forms the first deep gashes carved in the mountain slopes by eroding streams. Valiant protectors of less sturdy trees and plants, their whitened weather-sides bear witness to a fierce struggle for life on the bleak shoulders of the peaks.

Make your way, as the streamlets do, down to the alpine glades, on the high plateaus, where anemone, erythronium and calochortus push their buds through lingering snow-crusts. The scattered trees gather in their first groups.

Mazama Party resting among the sub-alpine firs in a flower-carpeted "park" at the foot of Mount St. Helens

A LoWland Ravine. Cedars, Vine Maples, Devil's Club and Ferns, near Mount St. Helens.

Just within their shelter pause for a moment. Vague distance is narrowed to a diminutive circle. The mystery of vastness passes. Sharp indeed is the division between storm-swept barren and forest shelter.

Here ravines, decked with heather, hold streams from the snowdrifts—streams that hunt the steepest descents, and glory in their leaps from rock to rock and from cliff to pool. If it be the springtime of the mountains—late July—the mossy rills will be half concealed beneath fragrant white azaleas that nod in the breezes blowing up with the ascending sun and down with the turn of day. Trailing over the rocks, or banked in the shelter of larger trees, creeping juniper (*Juniperus communis*), least of our evergreens, stays the drifting sands against the drive of winds or the wash of melting snows.

Along the streams and on sunny slopes and benches are the homes of the pointed firs. Seeking protection from the storm, the spire-like trees cluster in tiny groves, among which, like little bays of a lake, the grassy flowered meadows run in and out, sun-lit, and sweet with rivulets from the snows above. If you do not know these upland "parks," there is rare pleasure awaiting you. A hundred mountain blossoms work figures of white and red and orange and

The "Noble" Fir.

Dense Hemlock Forest, loWer West slope of Mount Hood.

blue in the soft tapestry of green. In such glades the hush is deep. Only the voice of a waterfall comes up from the canyon, or the whistle of a marmot, the call of the white-winged crows and the drone of insects break the stillness.

Mount Hood from Ghost-tree Ridge. Whitened trunks of trees killed by forest fires.

The outer rank of hemlock and fir droops its branches to the ground to break the tempest's attack. Within, silver or lovely fir (*Abies amabilis*) mingles with hardier forms. Its gray, mottled trunks are flecked with the yellow-green of lichen or festooned with wisps of moss

An Island of Color in the Forest. Rhododendrons and Squaw Grass on the west slope of Mount Hood.

"The common growth of mother-earth
Suffices me,—her tears, her mirth,
Her humblest mirth and tears."—Wordsworth.

Group of Red Cedars, five to eight feet in diameter.

down to the level of the big snows. And here, a vertical mile above the sea, you meet the daring western hemlock (*Tsuga heterophylla*), which braves the gale of ocean and mountain alike, indifferent to all but fire. It is of gentle birth yet humble spirit. It accepts all trees as neighbors. You meet it everywhere as you journey to the sea. But on the uplands only, in a narrow belt like a scarf thrown across the shoulders of the mountain, sub-alpine fir (*Abies lasiocarpa*) sends up its dark, attenuated spires, in striking contrast with the rounded crowns of its companions.

On the road to Government Camp, West of Mount Hood. Broadleaf Maple on extreme right; Douglas Firs arching the roadway, and White Fir on left.

A little lower, the transition zone offers a noteworthy intermingling of species. Down from the stormy heights come alpine trees to lock branches with types from warmer levels. Here you see lodgepole pine (*Pinus murrayana*), that wonderful restorer of waste places which sends forth countless tiny seedlings to cover fire-swept areas and lava fields with forerunners of a forest. Here, too, you will find western white pine (*Pinus monticola*), the fair lady of the genus, whose soft, delicate foliage, finely chiseled trunk, and golden brown cones denote its gentleness; and Engelmann spruce (*Picea Engelmannii*) of greener blue than any other, and hung with pendants of soft seed cones, saved from pilfering rodents by pungent, bristling needles.

Here also are western larch or tamarack (*Larix occidentalis*); or, rarely, on our northern peaks, Lyall's larch (*Larix Lyallii*), whose naked branches send out tiny fascicles of soft pale leaves; and Noble fir (*Abies nobilis*), stately, magnificent, proud of its supremacy over all. And you may come upon a rare cluster of Alaska cedar (*Chamæcyparis nootkatensis*), here at its southern

COPYRIGHT, WEISTER

Where man's a pygmy. A Noble Fir, 175 feet to first limb.

limit, reaching down from the Coast range of British Columbia almost to meet the Great sugar pines (*Pinus lambertiana*) which come up from the granite heights of the California sierra to play an important role in the southern Oregon forests.

Firs and Hemlocks, in Clarke County, Washington.

Across the roll of ridge and canyon, you see them all; and when you come to know them well, each form, each shade of green, though far away, will claim your recognition. Yonder, in a hollow of the hills, a cluster of blue-green heads is raised above the familiar color of the hemlocks. Cross to it, and stand amidst the crowning glory of Nature's art in building trees. About you rise columns of Noble firs, faultless in symmetry, straight as the line of sight, clean as granite shafts. Carry the picture with you; nowhere away from the forests of the Columbia can you look upon such perfect trees.

Westward of the Cascade summits the commercial forest of to-day extends down from an elevation of about 3,500 feet. Intercepted by these heights, the moisture-laden clouds are emptied on the crest of the range. Eastward, the effects of decreasing precipitation are shown both in species and in density. Tamarack, white fir and pines climb higher on these warmer slopes. Along the base of the mountains, and beyond low passes where strong west winds drive saturated clouds out over level reaches, western yellow pine (*Pinus ponderosa*) becomes almost the only tree. Over miles of level lava flow, along the upper

Deschutes, this species forms a great forest bounded on the east by rolling sage-brush plains that stretch southward to the Nevada deserts. Beyond the Deschutes drainage, where spurs of the Blue mountains rise to the levels of clouds and moisture, the forest again covers the hills, spreading far to the east until it disappears again in the broad, treeless valley of Snake river. North of the Columbia the story is the same. From the lower slopes of Mt. Adams great rolling bunch-grass downs and prairies reach far eastward. Here and there, over these drier stretches, stand single trees or clusters of western juniper (*Juniperus occidentalis*).

Fifty-year-old Hemlock growing on Cedar log. The latter, which was centuries old before it matured and fell, was still sound enough to yield many thousand shingles.

But on the west slope of the Cascades, and over the Coast range, the great forests spread in unbroken array, save where wide valleys have been cleared by man or hillsides stripped by fire. Here, in the land of warm sea winds and abundant moisture,the famous Douglas fir (*Pseudotsuga taxifolia*), Pacific red cedar (*Thuja plicata*) and tideland spruce (*Picea sitchensis*) attain their greatest development. These are the monarchs of the matchless Northwestern forests, to which the markets of the world are looking more and more as the lines of exhausted supply draw closer.

Sawyers preparing to "fall" a large Tideland Spruce.

Douglas fir recalls by its name one of the heroes of science, David Douglas, a Scotch naturalist who explored these forests nearly ninety years ago, and discovered not only this particular giant of the woods, but also the great sugar pine and many other fine trees and plants. As a pioneer botanist, searching the forest, Douglas presented a surprising spectacle

Sugar Pine, Douglas Fir, and Yellow Pine.

to the Indians. "The Man of Grass" they called him, when they came to understand that he was not bent on killing the fur-bearing animals for the profit to be had from their pelts.

The splendid conifer which woodsmen have called after him is one of the kings of all treeland. The most abundant species of the Northwest, it is also, commercially, the most important. Sometimes reaching a height of more than 250 feet, it grows in remarkably close stands, and covers vast areas with valuable timber that will keep the multiplying mills of Oregon and Washington sawing for generations. In the dense shade of the forests, it raises a straight and stalwart trunk, clear of limb for a hundred feet or more. On the older trees, its deeply furrowed bark is often a foot thick. Trees of eight feet diameter are at least three hundred years old, and rare ones, much larger, have been cut showing an age of more than five centuries.

To these areas of the greatest trees must come all who would know the real spirit of the forest, at once beneficent and ruthless. Here nature selects the fittest. The struggle for soil below and light above is relentless. The weakling, crowded and overshadowed, inevitably deepens the forest floor with its fallen trunk, adding to the humus that covers the lavas, and nourishing in its decay the more fortunate rival that has robbed it of life. Here, too, with the architectural splendor of the trees, one feels the truth of Bryant's familiar line:

The groves were God's first temples.

The stately evergreens raise their rugged crowns far toward the sky, arching gothic naves that vault high over the thick undergrowth of ferns and vine maples. In such scenes, it is easy to understand the woodsman's solace, of which Herbert Bashford tells in his "Song of the Forest Ranger:"

I would hear the wild rejoicing
Of the wind-blown cedar tree,
Hear the sturdy hemlock voicing
Ancient epics of the sea.
Forest aisles would I be winding,
Out beyond the gates of Care;
And in dim cathedrals finding
Silence at the shrine of Prayer.
* * * *
Come and learn the joy of living!
Come and you will understand
How the sun his gold is giving
With a great, impartial hand!
How the patient pine is climbing,
Year by year to gain the sky;
How the rill makes sweetest rhyming
Where the deepest shadows lie!

Fir, spruce and cedar you will see along the slopes of the Cascades in varying density and grandeur, from thickets of slender trees reclaiming fire-swept lands to broken ranks of patriarchs whose crowns have swayed before the storms of centuries. Among the foot hills, the pale gray "grand" or white firs (*Abies grandis*) rear their domes above the common plane in quest of light, occasionally attaining a height of 275 feet, while the lowly yew (*Taxus brevifolia*), of which the warrior of an earlier time fashioned his bow, overhangs the noisy streams. In the same habitat, where the little rivers debouch into the valleys, you may see the broad-leaf maple, Oregon ash, cottonwood, and a score of

Yellow Cedar, with young Silver Fir.

COPYRIGHT, GIFFORD

One of the Kings of Treeland —A Douglas Fir.

COPYRIGHT, JAS. WAGGENER, JR.

Firs and Vine Maples in Washington Forest.

lesser deciduous trees on which the filtered rays of sunshine play in softer tones.

Here and there in the Willamette valley you meet foothill yellow pine (*Pinus ponderosa var. benthamiana*), near relative of the western yellow pine. Oregon oak (*Quercus garryana*) occurs sparingly throughout the valleys, or reaches up the western foothills of the Willamette, until it meets the great unbroken forest of the Coast Range.

Towing a log raft out to sea, bound for the California markets.

The dense lower forests are never gaily decked, so little sunlight enters. But in early summer, back among the mountains, you may find tangles of half-prostrate rhododendron, from which, far as the eye can reach, the rose-pink gorgeous flowers give back the tints of sunshine and the iridescent hues of raindrops. Mingled with the flush of "laurel" blossoms are nodding plumes of creamy squaw grass, the beautiful xerophyllum. Often this queenly upland flower

A "Burn" on the slopes of Mount Hood, overgrown with Squaw Grass. Such fire-swept areas are quickly covered with mountain flowers, of which this beautiful cream-colored plume is one of the most familiar. Its roots yield a fiber used by the Indians in making baskets.

covers great areas, hiding the desolation wrought by forest fires. Its sheaves of fibrous rootstocks furnish the Indian women material for their basket-making; hence the most familiar of its many names. The varied green of huckleberry bushes is everywhere. They are the common ground cover.

In valley woodlands, the dogwood, here a tree of fair proportions, lights up the somber forest with round, white eyes that peer out through bursting leafbuds, early harbingers of summer. The first blush of color comes with the unfolding of the pink and red racemes of flowering wild currant. Later, sweet syringa fills the air with the breath of orange blossoms; and spirea, the Indian arrowwood, hangs its tassels among the forest trees or on the bushy hills. But the presence of deciduous trees and shrubs, as well as their beauty, is best known in autumn, when maples brighten the woods with yellow rays; when dogwood and vine maple paint the fire-scarred slopes a flaming red, and a host of other color-bearers stain the cliffs with rich tints of saffron and russet and brown.

A Noble Fir.

Western White Pine

Coming at last to the rim of the forest, you look out over the sea, where go lumber-laden ships to all the world. Close by the beach, dwarfed and distorted by winds of the ocean, and nourished by its fogs, north-coast pine (*Pinus contorta*) extends its prostrate forms over the cliffs and dunes of the shore, just as your first acquaintance, the white-bark pine, spreads over the dunes and ridges of the mountain. They are brothers of a noble race. You have traversed the wonder-forest of the world, and on your journey with the stream you may have come to know twenty-three species of cone-bearers, all indigenous to the Columbia country. Of these, one is Douglas fir, nowise a true fir but a combination of spruce and hemlock; seven are pines, four true firs, two spruces, two hemlocks, two tamaracks or larches, two cedars, two junipers, and the yew.

A Clatsop Forest. On extreme right is a Silver Fir, covered with moss; next are two fine Hemlocks, with Tideland Spruce on left.

So many large and valuable trees of so many varieties can be found nowhere else. A Douglas fir growing within the watershed of the Columbia is twelve feet and seven inches in diameter. A single stick 220 feet long and 39 inches in diameter at its base has been cut for a flagpole in Clatsop county. A spruce twenty feet in diameter has been measured. Such immense types are rare, yet in a day's tramp through the Columbia forests one may see many trees upwards of eight feet in diameter. One acre in the Cowlitz river watershed is said to bear twenty-two trees, each eight feet or more at its base. Though no exact measurements can be cited, it is likely that upon different single acres 400,000 feet, board measure, of standing timber may be found. And back among the Cascades, upon one forty-acre tract, are 9,000,000 feet—enough to build a town. Manufactured, this body of timber would be worth $135,000, of which about $100,000 would be paid to labor.

A Carpet of Firs; 300,000 feet, cut on one acre in a Columbia forest.

Along the Columbia you will hear shrill signals of the straining engines that haul these gigantic trees to the rafting grounds. Up and down the broad river ply steamboats trailing huge log-rafts to the mills. Each year the logging railroads push farther back among the mountains, to

Winter in the forest. Mount Hood seen from Government Camp road. Twenty feet of snow.

bring forth lumber for Australia, the Orient, South America, Europe and Africa. Many of our own states, which a few years ago boasted "inexhaustible" forests, now draw from this supply.

Since 1905 Washington has been the leading lumber-producing state of the Union, and Oregon has advanced, in one year, from ninth to fourth place. The 1910 production of lumber in these states was 6,182,125,000 feet, or 15.4 per cent. of the total output of the United States. The same states, it is estimated, have 936,800,000,000 feet of standing merchantable timber, or a third of the country's total.

This is the heritage which the centuries of forest life have bequeathed. Only the usufruct of it is rightfully ours. Even as legal owners, we are nevertheless but trustees of that which was here before the coming of our race, and which should be here in great quantity when our trails have led beyond the range. Our duty is plain. Let us uphold every effort to give meaning and power to the civil laws which say: "Thou shalt not burn;" to the moral laws which say: "Thou shalt not waste." Let us understand and support that spirit of conservation which

Rangers' Pony Trail in forest of Douglas and Silver Firs.

demands for coming generations the fullest measure of the riches we enjoy. For although the region of the Columbia is the home of the greatest trees, centuries must pass ere the seedlings of to-day will stand matured.

Forest Fire on east fork of Hood River. From a photograph taken at Cloud Cap Inn five minutes after the fire started.

Reforestation is indispensable as insurance. Let us see to it that the untillable hills shall ever bear these matchless forests, emerald settings for our snowpeaks. On their future depends, in great degree, the future of the Northwest. As protectors of the streams that nourish our valleys, and perennial treasuries of power for our industries, they are guarantors of life and well-being to the millions that will soon people the vast Columbia basin.

Reforestation—Three generations of young growth; Lodgepole Pine in foreground; Lodgepole and Tamarack thicket on ridge at right; Tamarack on skyline.

NOTES

Transportation Routes, Hotels, Guides, etc.—The trip from Portland to north side of Mount Hood is made by rail (Oregon-Washington Ry. & Nav. Co. from Union station) or boat (The Dalles, Portland & Astoria Nav. Co. from foot of Alder street) to Hood River, Ore. (66 miles), where automobiles are taken for Cloud Cap Inn. Fare, to Hood River, by rail, $1.90; by boat, $1.00. Auto fare, Hood River to the Inn, $5.00. Round trip, Portland to Inn and return, by rail, $12.50; by boat, $12.00. Board and room at Cloud Cap Inn, $5.00 a day, or $30.00 a week. Accommodations may be reserved at Travel Bureau, 69 Fifth street.

To Government Camp, south side of Mount Hood (56 miles), the trip is made by electric cars to Boring, Oregon, and thence by automobile. Cars of the Portland Railway, Light & Power Co., leave First and Alder streets for Boring (fare 40 cents), where they connect with automobiles (fare to Government Camp, $5.00). Board and room at Coalman's Government Camp hotel, $3.00 a day, or $18.00 a week.

Guides for the ascent of Mt. Hood, as well as for a variety of side trips, may be engaged at Cloud Cap Inn and Government Camp. For climbing parties, the charge is $5.00 per member.

The trip to Mount Adams is by Spokane, Portland & Seattle ("North Bank") Railway from North Bank station or by boat (as above) to White Salmon, Wash., connecting with automobile or stage for Guler or Glenwood. Fare to White Salmon by rail, $2.25; round trip, $3.25; fare by boat, $1.00. White Salmon to Guler, $3.00. Board and room at Chris. Guler's hotel at Guler P. O., near Trout Lake, $1.50 a day, or $9.00 a week. Similar rates to and at Glenwood. At either place, guides and horses may be engaged for the mountain trails (15 miles to the snow-line). Bargain in advance.

The south side of Mount St. Helens is reached by rail from Union station,. Portland, to Yacolt (fare $1.30) or Woodland ($1.00), where conveyances may be had for Peterson's ranch on Lewis River. To the north side, the best route is by rail to Castle Rock (fare, $1.90), and by vehicle thence to Spirit Lake. Regular guides for the mountain are not to be had, but the trails are well marked.

Automobile Roads.—Portland has many excellent roads leading out of the city, along the Columbia and the Willamette. One of the most attractive follows the south bank of the Columbia to Rooster Rock and Latourelle Falls (25 miles). As it is on the high bluffs for much of the distance, it commands extended views of the river in each direction, and of the snow-peaks east and north of the city. Return may be made via the Sandy River valley. This road is now being extended eastward from Latourelle Falls to connect with the road which is building westward from Hood River. When completed the highway will be one of the great scenic roads of the world.

From Portland, several roads through the near-by villages lead to a junction with the highway to Government Camp on the south side of Mount Hood (56 miles). The mountain portion of this is the old Barlow Road of the "immigrant" days in early Oregon, and is now a toll road. (Toll for vehicles, round trip, $2.50.) Supervisor T. H. Sherrard, of the Oregon National Forest Service, is now building a road from the west boundary of the national forest, at the junction of Zigzag and Sandy rivers, crossing Sandy canyon (see p. 71), following the Clear Fork of the Sandy to the summit of the Cascades, crossing the range by the lowest pass in the state (elevation, 3,300 feet), and continuing down Elk Creek and West Fork of Hood River to a junction with the road from Lost Lake into Hood River valley. The completion of this road through the forest reserve will open a return route from Hood River to the Government Camp road, through a mountain district of the greatest interest.

Southward from Portland, inviting roads along the Willamette lead to Oregon City, Salem, Eugene and Albany. From Portland westward, several good roads are available, leading along the Columbia or through Banks, Buxton and Mist to Astoria and the beach resorts south of that city. North of the Columbia (ferry to Vancouver), a route of great interest leads eastward along the Columbia to Washougal and the canyon of Washougal River (45 miles). From Vancouver northward a popular road follows the Columbia to Woodland and Kalama, and thence along the Cowlitz River to Castle Rock.

The tour book of the Portland Automobile Club, giving details of these and many other roads, may be had for $1.50 in paper covers, or $2.50 in leather.

Bibliography.—The geological story of the Cascade uptilt and the formation of the Columbia gorge is graphically told in *Condon: Oregon Geology* (Portland, J. K. Gill Co., 1910). For the Columbia from its sources to the sea, *Lyman: The Columbia River* (New York, G. P. Putnam's Sons, 1909) not only gives the best account of the river itself and its great basin but tells the Indian legends and outlines the period of discovery and settlement. *Irving: Astoria* and *Winthrop: The Canoe and the Saddle* are classics of the early Northwest. *Balch: Bridge of the Gods,* weaves the Indian myth of a natural bridge into a story of love and war.

The literature of the mountains described in this volume is mainly to be found in the publications of the mountain clubs, especially *Mazama* (Portland), *The Sierra Club Bulletin* (San Francisco) and *The Mountaineer* (Seattle). Many of their papers have scientific value as well as popular interest. It is to be hoped that the Mazamas will resume the publication of their annual.

Russell: Glaciers of N. Am. p. 67; *Emmons: Volcanoes of the U. S. Pacific Coast,* in *Bulletin of Am. Geog. Soc.,* v. 9, p. 31; *Sylvester: Is Mt. Hood Awakening?* in *Nat'l Geog. Mag.,* v. 19, p. 515, describe the glaciers of Mt. Hood. Prof. Reid has published valuable accounts of both Hood and Adams, with especial reference to their glaciers, in *Science,* n. s., v. 15, p. 906; *Bul. Geol. Soc. of Am.,* v. 13, p. 536, and *Zeitschrift fur Gletscherkunde,* v. 1, p. 113. An account of the volcanic activities of St. Helens by Lieut. C. P. Elliott, U. S. A., may be found in U. S. Geog. Mag., v. 8, pp. 226, and by J. S. Diller in Science, v. 9, p. 639.

The ice caves of the Mt. Adams district are described in *Balch: Glacieres, or Freezing Caverns,* which covers similar phenomena in many countries; by L. H. Wells, in *Pacific Monthly,* v. 13, p. 234; by R. W. Raymond, in *Overland Monthly,* v. 3, p. 421; by H. T. Finck in *Nation,* v. 57, p. 342.

Dryer's account of the first ascent of Mt. St. Helens may be found in *The Oregonian* of September 3, 1853, and his story of the first ascent of Mt. Hood in *The Oregonian,* August 19, 1854, and *Littell's Living Age,* v. 43, p. 321.

The Mountain Clubs.—For the following list of presidents and ascents of the Mazamas, I am indebted to Miss Gertrude Metcalfe, historian of the club:

	PRESIDENTS.	OFFICIAL ASCENTS.
1894	Will G. Steel	Mt. Hood, Oregon.
1895	Will G. Steel—L. L. Hawkins	Mt. Adams, Washington
1896	C. H. Sholes	Mt. Mazama (named for the Mazamas, 1896), Mt. McLoughlin (Pitt), Crater Lake, Oregon.
1897	Henry L. Pittock	Mt. Rainier, Washington.
1898	Hon. M. C. George	Mt. St. Helens, Washington.
1899	Will G. Steel	Mt. Sahale (named by the Mazamas, 1899), Lake Chelan, Wash.
1900	T. Brook White	Mt. Jefferson, Oregon.
1901	Mark O'Neill	Mt. Hood, Oregon.
1902	Mark O'Neill	Mt. Adams, Washington.
1903	R. L. Glisan	Three Sisters, Oregon.
1904	C. H. Sholes	Mt. Shasta, California.
1905	Judge H. H. Northup	Mt. Rainier, Washington.
1906	C. H. Sholes	Mt. Baker (Northeast side), Wash.
1907	C. H. Sholes	Mt. Jefferson, Oregon.
1908	C. H. Sholes	Mt. St. Helens, Washington.
1909	M. W. Gorman	Mt. Baker (Southwest side), and Shuksan, Washington.
1910	John A. Lee	Three Sisters, Oregon.
1911	H. H. Riddell	Glacier Peak, Lake Chelan, Wash.
1912	Edmund P. Sheldon	Mt. Hood, Oregon.

The organization and success of the Portland Snow Shoe Club are mainly due to the enthusiastic labors of its president, J. Wesley Ladd. Between 1901 and 1909, Mr. Ladd took a private party of his friends each winter for snow shoeing and other winter sports to Cloud Cap Inn or Government Camp. Three years ago it was determined to form a club and erect a house near Cloud Cap Inn. The club was duly incorporated and a permit obtained from the United States Forest Service. Mr. Ladd, who has been president of the club since its formation, writes me:

"Our club house was started in July, 1910, and was erected by Mr. Mark Weygandt, the worthy mountain guide who has conducted so many parties to the top of Mt. Hood. It is built of white fir logs, all selected there in the forest. I have been told in a letter from the Montreal Amateur Athletic Club of Montreal, Canada, that we have the most unique and up-to-date Snow Shoe Club building in the world. The site for the house was selected by Mr. Horace Mecklem and myself, who made a special trip up there. The building was finished in September, 1910. It is forty feet long and twenty four feet wide, with a six-foot fireplace and a large up-to-date cooking range. The organizers of the club are as follows: Harry L. Corbett, Elliott R. Corbett, David T. Honeyman, Walter B. Honeyman, Rodney L. Glisan, Dr. Herbert S. Nichols, Horace Mecklem, Brandt Wickersham, Jordan V. Zan, and myself."

The Portland Ski Club was organized six years ago, and has since made a trip to Government Camp in January or February of each year. The journey is made by vehicle until snow is gained on the foothills, at Rhododendron; the remaining ten miles are covered on skis. The presidents of the club have been: 1907, James A. Ambrose; 1908, George S. Luders; 1909, Howard H. Haskell; 1910, E. D. Jorgensen; 1911, G. R. Knight; 1912, John C. Cahalin.

The Mountaineers, a club organized in Seattle in 1907, made a noteworthy ascent of Mount Adams in 1911.

Climate.—The weather conditions in the lower Columbia River region are a standing invitation to outdoor life during a long and delightful summer. Western Oregon and Washington know no extremes of heat or cold at any time of the year. The statistics here given are from tables of the U. S. Weather Bureau, averaged for the period of government record:

Mean annual rainfall: Portland, 45.1 inches; The Dalles, 19 inches. Portland averages 164 days with .01 of an inch precipitation during the year, and The Dalles 74 days; but the long and comparatively dry summer is indicated by the fact that only 27 of these days at Portland and 15 at The Dalles fell in the summer months, June to September inclusive.

Mean annual temperature varies little between the east and west sides of the Cascades, Portland having a 57-year average of 52.8° as compared with 52.5° at The Dalles. But the range of temperature is greater in the interior. Thus the mean monthly temperature for January, the coldest month, is 38.7° at Portland and 32.6° at The Dalles, while for July, the hottest month, it is 67.3° at Portland and 72.6° at The Dalles.

While mountain weather must always be an uncertain quantity, that of the Northwestern snow-peaks is comparatively steady, owing to the dry summer of the lowlands. During July and August, the snow-storms of the Alps are almost unknown here. After the middle of September, however, when the rains have begun, a visitor to the snow-line is liable to encounter weather very like that recorded by a belated tourist at Zermatt:

First it rained and then it blew,
And then it friz and then it snew,
And then it fogged and then it thew;
And very shortly after then
It blew and friz and snew again.

Erratum.—On page 72, I have been misled by Dryer's statement into crediting the first ascent of Mount Hood to Captin Samuel K. Barlow, the road builder. The mountain climber was his son, William Barlow, as I am informed by Mr. George H. Himes, of the Oregon Historical Society.

INDEX

Figures in light face type refer to the text, those in heavier type to illustrations.

Klickitat River Canyon, near Mount Adams.

Spotted something that isn't quite right?

Let us know and we'll fix it for future reprints.
And as a thank you for your diligence, we'll
send you a free Sothis book of your choice.

publisher@sothis.press

This page intentionally left blank

Made in United States
North Haven, CT
05 November 2026

21817322R00088